American Dreamer

Bucky Fuller & the Sacred Geometry of Nature

In memoriam

Berenice Jacqueline Eastham
– 1923-2004 –

Magna Mater

American Dreamer

Bucky Fuller & the Sacred Geometry of Nature

Scott Eastham

The Lutterworth Press

The Lutterworth Press
P.O. Box 60
Cambridge
CB1 2NT
United Kingdom

www.lutterworth.com
publishing@lutterworth.com

ISBN (13): 978 07188 3031 1

British Library Cataloguing in Publication Data
A record is available from the British Library

First Published in 2007

Printed in the United Kingdom by
Antony Rowe, Chippenham

Contents

Acknowledgments

First and foremost, of course, thanks are due to Bucky Fuller himself – not just for the inspiration to undertake this book, but for his permission to reprint figures from *Synergetics*. Next, my sincere appreciation goes to his collaborator Ed Applewhite, who gave me the chance to meet Fuller and offered valuable suggestions for the project.

Illustrations from R. Marx & R. B. Fuller, *The Dymaxion World of Buckminster Fuller* as well as J. Krause & C. Lichtenstein, *Your Private Sky*, are reproduced here with the permission of the Estate of R. Buckminster Fuller; the captioned isotropic vector matrix on page 31 appears by permission of the William Hugh Kenner Estate.

The striking black-and-white photos which accompany the early parts of this book were provided by Robert Duchesnay, whose Dome Project has over many years reminded the city and residents of Montréal that their Expo Dome is an important artifact worth preserving. Duchesnay's photos have an enigmatic quality which may have helped provoke me to take the unusual approach to Fuller's legacy – via the American dream, New England transcendentalists, *anima mundi* and mandalas, etc. – which emerges in these pages.

I first met Robert as well as many other Fuller afficionados when I had the good fortune to conduct the Fellows Seminars at Lonergan College (Concordia University) in Montréal during the academic year 1987-88. I would like to thank Mark Doughty for inviting me to contribute to that ongoing series of 'major figures' lectures. I recall with great relish the many lively hours of debate and discussion in the company of both fellows and students. Such a setting and such a dedicated group of scholars is a rare find in today's academic world. Charles Davis, then Principal of the College, has since passed away, but his open scepticism about Fuller's ideas and his probing, rationalist critiques may well have elicited some of the arguments offered here.

I cannot recollect my early days working on Fuller without bringing to mind my fellow explorer John Blackman – co-author of the Workbook/Appendix *Unfolding Wholes* here – who made those studies a daily delight. John's good sense, good humour and astute intuition led us both down byways I might never have explored on my own. In those days, we were often abetted by Michael Connolly, who proved a dab hand at 'unfolding' Wholes himself, and Tom Parker, who skillfully rendered such figures in accurate line drawings. Diane Pendola provided some surprising support as well as 'space' for things to happen, while Kathy Borst no doubt understood Fuller's math and its pedagogic possibilities better than any of us. All of these lively enthusiasts let us know in different ways that we were onto something worthwhile.

My thanks also go to the Lutterworth staff: Amy Dampney for her painstaking copy-editing, Tom Wilson for book design, Emma Collison and Aidan Van de Weyer for help with the graphics, and of course Adrian Brink, who committed his house to the project in the first place and stuck with it through thick and thin.

Finally, I should thank my family for putting up with this long-lived and sometimes demanding project. My wife, Mary, has been especially tolerant, but my daughters Casey and Alison also got to know 'Uncle Bucky' through those funny geodesic models that kept turning up in their childhood years everywhere from cribs to Christmas trees.

So to all of these – friends, family, colleagues – I offer this book, along with my thanks.

Introduction

An Imaginary Symposium

It would be easy to make R. Buckminster Fuller an early apostle of globalization. Indeed, his cherished word 'synergy' has lately been so successfully co-opted by financiers and corporations to rationalize so many callous strategies for global profiteering that it may well qualify as one of the most abused words of the late twentieth century. But I am going to take another tack. The 1999 'Battle of Seattle' WTO protest as well as the more recent Washington D.C., Prague, Genoa, Barcelona and Cancun protests against the G8, the World Bank and the IMF, show us that stark battle lines between globalism and local values of all sorts – economic, ecological, civic, cultural, and religious values – have already been drawn in our midst. With the fighting factions forming, on which side of the police line would Bucky Fuller stand today? The short answer is that he would not approve of the way 'globalization' is going today, any more than he sanctioned corporate greed in his lifetime.

By the same token, it would be entirely straightforward to treat Fuller as a prophet of technological development, or at least as a prodigious inventor of artifacts, and to turn this book into yet another catalogue of his marvelous technological innovations – the geodesic dome, the Dymaxion 'Omni-Medium Transport,' the self-contained bathroom, and so on. To ignore cultural context, to assume that technical facts lie somehow outside the bounds of cultural critique, is a common American blindspot. Yet no technology is neutral. Each new technology is already inflected in a myriad of ways by the culture which conceives it, and in turn may drastically alter all the variables of the delicate balancing act between nature and culture. Fuller knew this, and in fact designed his inventions with the specific aim of bringing about far-reaching cultural transformations.

In other words, there are plenty of one-sided ways and partial perspectives with which to treat Fuller – inadequately. What then would be an adequate approach, a holistic approach, to a man who espoused "the exploratory strategy of starting with the whole"?

Imagine the ideal symposium on Buckminster Fuller. You would need a biographer to fill in some of the details of his busy life; an historian to 'place' him in his American context; an architect or structural engineer to explicate the principles with which he built things; a non-dogmatic mathematician to outline his synergetic geometry; a philosopher to examine the assumptions of his often unsettling pronouncements; a theologian, perhaps, to recognize the persistent threads of spirituality running through his work; a social critic to weigh his claims that Western society has organized itself on specious premises; a poet or literary critic to do justice to the images he presents in his "mental mouthfuls"; and maybe an artist or computer animator to render it all accessible in engaging diagrams.

Nobody can be all of these things, and yet each of us must perforce respond to Bucky Fuller in multiplex ways. Just as no single perspective will adequately render one of his geometric figures, Fuller resists being 'reduced' to only one or two of his many facets. You may consider this book to be something like the 'proceedings' of such an imaginary symposium. I have tried to present here multiple perspectives, and at least to sketch out the various domains in which Fuller's multifarious thought and work are pertinent. There are bound to be clashes between such diverse views of Fuller, which cannot just be papered over for consistency's sake. Both technophiles and technophobes, for instance, have long squabbled over the meaning of Bucky's legacy, and a single book is unlikely to resolve their dispute – though it might hope to deepen *both* perspectives, so that the argument begins to turn into a dialogue. But knowledge is always personal knowledge, as Michael Polanyi used to insist. My approach to Bucky will, therefore, reflect my own interests, too, my own peripatetic readings, and the particular Fuller constructions I have personally had a chance to visit.

Indeed, this book represents for me a kind of intellectual odyssey, a return to meaningful origins. It started out as an attempt to assess the legacy of Buckminster Fuller, scientist-artist-engineer extraordinaire. It passed through a series of discoveries about the original American dream informing Fuller's vision, which I hold to be an ethic of cooperation, not competition. Naturally enough, my family's move in 1993 from North America to New Zealand prompted me to re-examine just which dimensions of the American dream were specific to place, and which were portable. The whole project ended up, a little surprisingly, in what I can only call a rediscovery of the world soul, the 'omnitriangulated' structure of a living Universe in Fuller's work, which I believe revives the *anima mundi* beloved of the ancients. If you want to save the world, it seems clear to me now, you first have to save her soul …

In other words, my approach, like anybody else's, is bound to be idiosyncratic. To claim 'objectivity' in such a study would be obtuse and misleading, just as Bucky Fuller's science was creative and individualistic in the extreme, and would not fare well in today's climate of 'team science' in thrall to big research grants from governments and transnational corporations. Bucky went his own way, and discovered things which a more conventional methodology would not have permitted him to find. Practically the last words of the last lecture I heard him deliver were to follow your own sense of what is true, and allow yourself to be corrected by the Universe. I have tried to adopt at least that much of his method.

American Dreamer does not, therefore, always follow a strictly linear scheme of exposition, but often takes a circular or even spiralling path. Beginning in the ruins of the Montréal Expo Dome, Bucky's 'Taj Mahal' and a center of gravity for this book, various lenses for focusing facets of Fuller's work radiate outward in ever-widening arcs – first into the New England region and the 'spirit of place' to be found there; next examining the vicissitudes of the American dream at large; then using the full range of Fuller's artifacts to raise questions of technology and human values which eventually span the globe; and finally landing back in the moral and spiritual climate of New England. These concentric probes of context open out in time as well; like deep sea soundings, they are attempts to fathom the various pasts and futures which inform our ever precarious present.

I make no apologies for 'lateral' thinking. If the book works at all, it will cohere in a non-linear way, like the tensional great circle continuities which hold together the figures of Bucky's synergetic geometry. The various intersecting themes will appear to slice cleanly into one another, sometimes at abrupt and unexpected angles. This is my intention, a kind of structural homage to Fuller. These underlying continuities seem to me a crucial dimension of his legacy. I can for instance never long ignore the voice of Ralph Waldo Emerson as at least one of Fuller's intellectual progenitors – "The mind of Emerson," asserts Harold Bloom, a little imperiously, "is the mind of America" – but I can also hear echoes from the Native American context, and register even more distant reverberations in both the iconographies and the languages of the Indo-European heritage.

This raises another way to look at this book: It is a conversation with ghosts. As a Californian visiting New England in search of Fuller's roots, I found myself in another country. California may not be 'Ecotopia,' but it is also surely not the old homestead of early Americana. New England felt to me a little like time travel: Emerson, the Shakers, the pilgrims and American founding fathers, as well as the ghostly voices that first called together the Iroquois Confederacy – all came alive for me, and spoke to me of another dream behind today's highly touted American dream of material success. They still do, and their persistent presence underscores the strange fate of that original American dream.

The writing of this book was catalyzed not only by Fuller's 1995 Centennial, which ironically coincided with that of his arch-critic Lewis Mumford, but also by Robert Duchesnay's painstaking efforts to photographically document the remains of Fuller's large constructions. What I see in the arresting Duchesnay photographs – beautiful structures often falling to bits – is mainly the interplay of syntropy and entropy. In closed systems, entropy is the tendency of things to lose energy and fall apart. Syntropy, by contrast, is Fuller's word for the cohering forces that hold things together in an open, dynamic universe of living energies. In the physical world, entropy affects everything – including Fuller's Expo dome, which I first visited in ruins in 1987. In Fuller's universe, though, the integrative (syntropic) forces are stronger than entropy. He claims the integrity of life survives the all-too-evident powers of decay and disintegration. I wanted to see if he was right, so I looked to his own ideas and structures to see how they had weathered the ravages of time.

Thanks to the good offices of his collaborator Ed Applewhite, I was fortunate enough to spend several hours of bracing conversation with Fuller towards the end of his life. I also had the chance to correspond at some length with Lewis Mumford. Long after their passing, their voices continued vigorously to dispute one another – point, counterpoint; endless, insistent echoes – in my head. I knew of their deaths, of course, but it was only in the writing of this book that I came to appreciate the vitality of their 'ghosts': strong, clear voices importuning me on either hand, sometimes seeming to speak right past me to one another. So one purpose of this book is not so much to lay those ghosts to rest, as to pass on to the reader some hints of their animated conversation about the meaning and ends of human life in a technological society.

Finally, a word about the long Appendix/Workbook entitled "Unfolding Wholes."

Stanley Cavell maintains that Ralph Waldo Emerson was the first to discover America *in thought*. Much the same might be said for Bucky Fuller discovering how to *think* about planet Earth. His *Synergetics* is subtitled "Explorations in the Geometry of Thinking." Yet although his thinking presents itself in scientific terms, Fuller did not turn his insights over to peer-reviewed journals for acceptance by the mainstream scientific community. That (partial) acceptance has come instead through quite unexpected applications of his thinking – Buckyballs (Carbon 60) would be the most famous instance, or his seemingly far-out 'Polynesian Genesis' hypothesis which acquired a measure of plausibility when paleoanthropologists discovered *homo erectus* remains more than 1.5 million years old in Indonesia. But Bucky's 'method' was simply to rely on his own experience, and to ask his audiences to do likewise. Whenever he visited a college or university in the 1960s, the students started building geodesic domes with him; the 'proof' was in the 'putting' of the principles to work. My aim here is more modest, namely to provide the reader with specifications for seven Fuller figures which may be constructed with ordinary household materials. If you want a demonstration of Fuller's principles, so that you may see for yourself whether they work as he says they do, then these 'local globes' will serve you well. Before his death, Fuller was kind enough to grant permission to reproduce specifications for these figures from *Synergetics*.

Because there is always more to be said, the best introduction to this book was probably penned long ago by Emerson in the opening lines of his "Circles," a passage often taken to be the purest statement of American transcendentalism:

> *The eye is the first circle; the horizon which it forms is the second; and throughout nature this primary figure is repeated without end. It is the highest emblem in the cipher of the world. St. Augustine described the nature of God as a circle whose centre was everywhere, and its circumference nowhere.*[1] *We are all our lifetime reading the copious sense of this first of forms … Our life is an apprenticeship to the truth, that around every circle another can be drawn; that there is no end in nature, but every end is a beginning; that there is always another dawn risen on mid-noon, and under every deep a lower deep opens.*[2]

SE

Montréal Expo Dome (© Robert Duchesnay,1988.)

I. Archaeology of a Vision

1987. We come late upon the scene. A tremendous shape mushrooms up before us, as indifferent to our gaze as it is to its frozen surroundings. It's as if something once happened here, some cosmic drama played upon this stage, and we are left only the quiet shell of that event. The forces once in play here seem to have played themselves out. We bundle our overcoats to us and huddle before yet another bitter blast of the Montréal winter wind.

This thing, this bare Expo Dome, seems an alien form here. It stands out and it stands totally alone, somewhere – or when – it shouldn't be. It is an *American* thing, out of place; unmoved by the deep Québec cold, by the iced-up St. Lawrence River, by the sharp-edged skyline of Montréal above and beyond the ice. The dome seems to hold its own like an architectural island on its own island, Ile St. Hélène, a shape that has kept its shape despite all the elements – fire, rain, snow, ice and wind – that time and disuse have hurled against it.

We move closer to the strange thing, into its black lace shadow. We circle round, then duck under the protecting fence through a low furrow made by many 'illegal' visitors over the years. The tourists and architecture buffs, the Bucky Fuller freaks, the curious, the daring or the merely foolhardy, the teenaged beer drinkers and furtive lovers and grafitti artists; many have made this pilgrimage before us, too many leaving their rubbish. We enter the dome compound cautiously, keeping an eye on the broken chunks of concrete, the twisted lengths of metal beams and jagged protusions everywhere underfoot.

My guide, Robert Duchesnay, knows every nook and cranny of this ruin. To muse over ruins is customarily considered a Romantic pursuit, but for that sensibility Nature itself was the idyll. You made a word-picture, or painted a landscape, and to it you added an artificial ruin to heighten the sense of Nature's dominance over Man's puny handiwork. But the Expo Dome plays on this sensibility in an unusual way. It is not the past, or Nature in the raw, that is lost or to be found in this place. Rather it is the *romance of the future* that draws people here, the glimpse of a future that once looked bright and hopeful, and now has grown dim and darkly clouded. It evokes nostalgia, yes, but not for the past or the pastoral scene. What we feel instead is something like nostalgia for a lost future: *Once upon a might-have-been …*

Anyway, this particular 'ruin' stoutly defies decrepitude. It has been called 'the most beautiful building-completely-destroyed-by-fire in the world.' Some, including its architect, Buckminster Fuller, have even declared that they prefer the ruin to the once-functional building. And it's easy to see why.

The structure itself is so elegant; the vast, mathematically precise sphere so overwhelming; the feel of the thing at once so massive and so lightsome, that it defies being taken in at a single glance. Literary critic Hugh Kenner once noted that it looks much smaller from the outside, since on a sphere (or two-thirds sphere, in this case) the external dimensions retreat from the eye. We're not used to looking at buildings like this; it is outside our usual 'frameworks.' Inside, it is clearly huge – very clearly, since almost nothing stands between you and the enormous 20-storey vault overhead.

For only this much of his achievement was Fuller grudgingly honored by the American Institute of Architects, that he enclosed more space with less stuff than anybody else in human history. Although he eventually built even larger structures, the Expo Dome still dwarfs the Pantheon, St. Peter's, the Taj Mahal, the Hagia Sofia, and all history's other famous domes. Today it simply stands there – vast, purposeless, and yet somehow its own justification. Filmmakers have repeatedly been drawn to it: Robert Altman shot his frigidly futurist *Quintet* under it in 1978; it appears in passing in Stuart Cooper's moody, mechanically symmetrical thriller, *The Disappearance.* The films glance toward the image, but say not a word about it.

2005. Montréal, which inherited the Dome after Expo '67, has never quite figured out what to do with it. Slow-growing pressure at last prompted some clean-up and refurbishing in the early 1990s. In 1995, the Centennial of Fuller's birth, the Dome was grandly rechristened 'Biosphere' and turned into an environmental education center. It contains a facility for monitoring water-quality in the St. Lawrence River, standing lonely sentinel-duty for science while commerce and industry continue to poison the Seaway. Despite Montréal's lack of interest, the Dome has, over the years, become something of a symbol for the city. For many of us, I suspect, it also serves as a kind of symbol for the 1960s, a lost era of hope which seemed to peak in 1967 when the Mohawk high-steel workers were erecting the Expo Dome – *without scaffolding*, believe it or not – and the Summer of Love was about to get off the ground in San Francisco. The romance of the future still had power then. People were actually getting ready to walk on the moon. Optimism and energy coupled with science and engineering seemed as if they might very well guarantee a limitless horizon for human 'progress.'

All that is gone now, of course.[3] But if we are to understand this strange webwork, this time capsule to futures past, we must take account of the way in which it came to be: first, spun in Fuller's own mind; then built as the United States Pavillion for the 1967 Expo; later destroyed through negligence and misadventure; and finally, left standing ever since as a kind of open enigma, a challenge to the imaginations of all who happen upon it.

Yankee Ingenuity

R. Buckminster Fuller (1895-1983) has been characterized as a mathematician, architect, inventor, philosopher, and even a prophet. He has also been caricatured as a wide-eyed technocrat with big plans for the industrialization of practically everything. Toward the end of his life, a listing of his various honorary degrees made his the longest entry in *Who's Who.* All of these classifications notwithstanding, Fuller tended to refer to himself as a "comprehensive anticipatory designer,"[4] which was his own way of summing up five decades of unparalleled Yankee ingenuity.[5]

It is also fair to understand 'Bucky' Fuller, as he liked to be called, as a poet or an artist – that is, a maker of images.[6] And he was a poet not mainly for the volleys of verse he took to publishing in his later years, each line a "mental mouthful" as he put it, but rather because the great part of his life's work was to propose what may well be the grandest pattern metaphor of our time: *Nature's coordinate system,* a triangulated matrix which has turned out to be an astonishingly accurate model for natural structures – like the posthumously-named 'Buckminsterfullerene' (C_{60}), *Science* magazine's 1991 "Molecule of the Year."[7]

Fuller called his geometry "synergetics," or the "geometry of thinking." Its basic approach is "the exploratory strategy of starting with the whole ..." From such thinking about housing in an integral way, for example, he came up with the startling innovations for which he is justly famous – geodesic domes and the 'Dymaxion Dwelling Machine' most notable among them. Fuller's geometry is founded upon triangles and tetrahedra – as is all the carbon chemistry of organic life – and leaves behind the square, cubical and gridiron forms which still saddle most conventional architecture. Indeed, Fuller's synergetics goes well beyond architecture altogether, into the very structure of matter and the shape-shiftings of the cosmos at large. So it helps to consider his "geometry of thinking" not as some abstract mathematical formalism, but as an artistic exploration, an adventure of the creative imagination. No one ever took Einstein's relativity more to heart than Fuller; everything is constitutively related to everything else. Bucky Fuller's achievement was to articulate these very basic relationships of matter and energy in meticulous detail, and to find practical applications everywhere, most of which are still well ahead of their time. His innovations span so many fields of science and engineering so adroitly that Marshall McLuhan may not have been too far off the mark in dubbing Fuller "the Leonardo da Vinci of our time."[8]

Over the course of 1995, those who knew him or were touched by him in life celebrated the Centennial of Bucky's birth. Since then, the nostalgic Goodman/Simon *American Masters* PBS production *Thinking Out Loud* has offered rare and touching glimpses of both the public and private man from half a century of documentary footage, and J. Baldwin has commenced in his *Bucky Works* a long overdue reexamination of the continuing utility of Fuller's many technical innovations for coming to grips with present-day ecological and economic dilemmas. But lacking the magic of Bucky's personality, his legend has otherwise quickly faded from the media spotlight. Architecture critic Allan Temko is probably right to call the Expo Dome the apogee of Fuller's career; in 1967, Bucky was incontestably the most famous structural designer in the world. An entire generation has since come of age who have never felt for themselves the magnetic influence of Buckminster Fuller, or tested the strength his ideas against their own experience.

The best chance for newcomers to 'meet' Bucky has probably been provided by D. W. Jacob's remarkably comprehensive one-man play, "R. Buckminster Fuller: The History (and Mystery) of the Universe," performed memorably by Ron Campbell first at the San Diego Repertory Theater in the Spring of 2000, and then to multiple return engagements in San Francisco and elsewhere ever since. Next best would have to be the very extensive traveling museum exhibit, *Your Private Sky,* mounted by the Zürich Museum für Gestaltung which toured Europe in 2000 and Asia in 2001. The vast photographic tome compiled from this exhibition by Joachim Krause and Claude Lichtenstein delves deeply back into the Fuller archives to bring to light many previously unremarked aspects of Fuller's early formative years.[9]

Of course the past decade has divulged a rich harvest of fresh and sometimes astonishing discoveries associated with Carbon 60 and its molecular family, the 'fullerenes.' In April 1996, for instance, marine geochemists Luann Becker and Jeffrey Bada detailed in *Science* their findings that naturally occurring 'Buckyballs' – hollow soccer-balls of soot, each made of 60 or 70 carbon atoms – were created by a meteor the size of Mt. Everest crashing into the Ontario region nearly two billion years ago, and appear to contain gases from a distant star that expired long before our own sun even ignited – making these the oldest complex molecules so far found on Earth.[10] Fuller would have been delighted by such findings. There are no native sources of carbon on planet Earth, yet all organic lifeforms are carbon-based. Where did it (and eventually all of us) come from? We are all made of star-stuff, Bucky used to say; in light of today's evidence, he might have added that the death of such stars (particularly in their red giant phase, when main sequence stars like our Sun become prolific producers of carbon) may well have helped kick-start the carbon chemistry of life on this planet. But such new findings reflect only the most indirect credit on their namesake.

What is the measure of a man? His ideas and accomplishments? Of course. His friends, followers, influence? Yes, all of these. But above all, and in a sense *beneath* all these there is something else – a character, a style of thinking, a certain spirit, a way of being. In this respect, Buckminster Fuller was a man distinguished by the depth of his determination to think for himself, to resolutely go his own way in both words and works. In 1927, at about age 33, Fuller decided to stop speaking until he could rid himself of received opinions, untested assumptions and speech patterns not his own. When he resumed speaking a year later, the results were altogether unusual. Thinking, he had decided, meant weeding out irrelevancies. Henceforward, he would speak only of matters he had found to be true in his own experience, or of things which could be tested by experiment. Not surprisingly, both his new way of speaking and the style of thought behind it were unique and, to many people, quite incomprehensible.

For Fuller, the sun does not rise or set, there is no up and no down (on our spheroidal Earth), and there are no solids or straight lines in all Universe. If you had just heard or read about such assertions, which aim to carry Einstein's postulates to their logical term, you might well come away scratching your head. But when Fuller began to *show* people what he was talking and thinking about by building elegant models – Try it! Hands on! See for yourself! – his approach became strikingly clear. Buckminster Fuller had hit upon an entirely novel way not only of expressing himself, but of relating to the Universe at large. Indeed, he devoted his life to finding ways for people to work *with* Nature's principles, rather than building, dwelling and thinking against or in spite of the patterned dynamisms of the natural world. On his deathbed Albert Einstein said he found himself facing Ernest Haeckel's famous question, which may yet turn out to be crucial for our own ecologically-stricken era: "Is the universe friendly?" In 1971, Ezra Pound called Bucky Fuller "friend of the universe," an epithet which suggests that here at least is a man whose life and work offers a resounding "Yes!" to that query.[11]

Fuller saw in technology the human mind at work finding ways to do "more and more with less and less," using Nature's bounty without abusing her limits. Early on, he made a distinction between 'Brain,' which he saw mainly as a mechanism for storing particular, special-case experiences and 'Mind,' which he saw as the capacity to learn, to discover the generalized principles which connect such experiences and make them meaningful. 'Brain' is physical and local; 'Mind' metaphysical and, he would insist, universal. Just so, he came to consider all of his inventions and buildings as no more than models, graphic demonstrations of what could be done by a mind in tune with Nature. And for those of us who come after Fuller, his works offer the possibility of an alternative science, a way of exploring natural and human possibilities for 'livingry,' as he called it, rather than weaponry.

With Fuller's models in their hands, Einstein's universe is intuitively and spontaneously comprehensible by children. All children are born geniuses, he used to say, until parents and schools teach them to distrust their own innate sense of the truth. And it is in those children that Fuller's hope for the future resided, perhaps because he managed to preserve in his own way of looking at things so many of those richly creative characteristics we associate with children: unflagging curiosity coupled with delight in the unknown and unexpected. The motto for his magnum opus, *Synergetics*, was "Dare to be naive." Not the whole picture, to be sure, but an important aspect of a man who claimed that he was not a genius, but rather only an average human being with "a terrific bundle of experiences."

American Dreamers

There are many ways of assessing Bucky Fuller's legacy. But picking about in the charred ruins of the once-glittering American Pavilion for Expo '67, it doesn't take long to realize one is also picking one's way through some other remnants as well, torn cast-offs from the very fabric of the American dream itself. To see how deeply the texture of Fuller's own vision is woven into the many-layered historical tapestry of the American dream, we must first try to sort the living strands from tangled skein of lost hopes and shredded dreams it has become, and then try to stitch those strands together in some semblance of the original pattern. Betsy Ross had it easy ...

Once there really was an American dream. No, not the materialistic one with two of everything: two kids in the big house, two cars in the garage, two pets in the yard, two bank accounts, and maybe two marriages before you get it all right. The original American dream was part of the spirit of this place, this land, this Turtle Island as the Iroqouis call North America. Its homeplace was not the Great Plains or the desert Southwest or the Rocky Mountains or the California goldfields. It belonged to New England, or to be precise, to the Northeast woodlands more or less centered on today's New England. It was alive there when the white settlers first arrived from England, it was kept alive for a couple of hundred years by remarkable individuals and even more remarkable communities and utopian experiments. Its spirit was that of a bold, pragmatic, reflective, self-reliant people who found themselves in a rich land and attuned themselves to its possibilities for new sorts of human dwelling. It has perhaps never been quite so clearly articulated as by Ralph Waldo Emerson:

> *We have listened too long to the courtly muses of Europe. The spirit of the American freeman is already suspected to be timid, imitative, tame ... What is the remedy? ... If the single man plant himself indomitably on his instincts, and there abide, the huge world will come round to him. Patience, – patience; with the shades of all the good and great for company; and for solace the perspective of your own infinite life; and for work the study and communication of*

principles, the making those instincts prevalent, the conversion of the world ... We will walk on our own feet; we will work with our own hands; we will speak our own minds. A nation of men will for the first time exist, because each believes himself inspired by the Divine Soul which also inspires all men.[13]

"Life alone avails, not the having lived," declared Emerson, casting off the shell of European conventions as he sought a new birth of the soul appropriate to a new land. As a thinker, it is entirely appropriate to situate Fuller within the tradition of American transcendentalism best exemplified by Emerson, who happens to have been a close friend of Fuller's famous great-aunt, Margaret Fuller. He himself referred to her glittering nineteenth-century circle of New England literati as "the small coterie of thinkers who formed the original nucleus of an American culture."[14] With all the optimism of his transcendentalist predecessors, Bucky would steadfastly maintain that humans were meant to be a success on this planet – and all the apparent evidence in the world to the contrary would never dissuade him from this conviction; our immediate prospects, in his phrase, were "utopia, or oblivion." Work against Nature's principles, he would say, and you will find yourself thwarted at every turn. Work *with* Nature's principles, and the Universe itself pitches in to help. There is at work in every human endeavor a larger mind, something like the very soul of Nature. Harking back to this ancient tradition of the living Universe or *anima mundi,* Emerson spoke of the 'Over-Soul,' and Fuller of the intellectual integrity of Universe. The following passage is vintage Emerson, but presages Fuller in both spirit and substance:

Beauty rests on necessities. The line of beauty is the result of perfect economy. The cell of the bee is built at that angle which gives the most strength with the least wax; the bone or quill of the bird gives the most alar strength. 'It is the purgation of superfluities,' said Michel Angelo. There is not a particle to spare in natural structures.[15]

Like Emerson, Fuller chose to rely primarily on his own experiences in order to discover the intimate secrets and principles by which the Universe cohered. He once wrote:

I did not set out to design a house that hung from a pole or to manufacture a new type of automobile, invent a new system of map projection, develop geodesic domes or Energetic Geometry. I started with the Universe – as an organization of regenerative principles frequently manifest as energy systems of which all our experiences, and possible experiences, are only local instances. I could have ended up with a pair of flying slippers.

Here the analogy with Leonardo da Vinci is just about irresistible.[16] Lacking (and professing to scorn) the classical book-learning that ruled the academy of his time, the West's first supremely inventive "disciple of experience" also resolved to put himself directly under the tutelage of Nature, for reasons similar to Bucky's:

Although human ingenuity makes various inventions, corresponding by various machines to the same end, it will never discover any inventions more beautiful, more appropriate or more direct than nature, because in her inventions nothing is lacking and nothing superfluous.[17]

As for Bucky, there seems to be something in the New England air that fosters a non-conformist individuality; it's practically an indigenous tradition. Fuller was raised in this atmosphere, in Massachusetts, and spent much time in later life on his family estate on Bear Island, Maine. However radical his ideas, he always dressed like a strait-laced New England Protestant minister making his rounds – of 'Spaceship Earth.'[18] He imbibed his independent spirit early, even managing the dubious distinction of being the only person ever to be expelled from Harvard University *twice.* (In his last years, he held the Charles Eliot Norton Chair of Poetry at Harvard, perhaps a unique experience for a Harvard maverick.) But the complex of notions that guided his work – that nature is the best teacher, that technology moves toward a refined simplicity of life, that industry should meet basic human needs like housing, and that such pragmatic concerns are not in the least divorced from spirituality – seems to be basic to the New England temperament.

From the early seventeenth century onward, villages in New England and environs exhibited what has sometimes been called 'Yankee communism,' a spirit of corporate co-partnership carried over to this day in communal barn-raisings amongst the Pennsylvania Dutch, for example. It did not escape the eye of the ever-perspicacious Alexis de Tocqueville, who observed in 1831:

> *The Americans ... show with complacency how an enlightened regard for themselves constantly prompts them to assist one another and inclines them willingly to sacrifice a portion of their time and property to the welfare of the state.*[19] *... The free institutions which the inhabitants of the United States possess, and the political rights of which they make so much use, remind every citizen ... that he lives in society. They every instant impress upon his mind the notion that it is the duty as well as the interest of men to make themselves useful to their fellow creatures.*[20]

Citing such passages, Hector Garcia writes in the present day of the urgent need for America to rediscover this sense of 'cultural complementarity': "It seems that the origin of American excellence might have been more about cooperation than competition."[21] Six decades earlier, Dartmouth historian and jurist Eugen Rosenstock-Huessy had seen just this: "'Co-operate' is the most striking phrase of the American vocabulary. For concrete co-operation, not for abstract philosophy, reason was given to men ... [the] co-operative reasoning of the men of good will."[22] Unfortunately, for America as for human nature generally, this communitarian spirit seems more likely to emerge in times of disaster – in the aftermath of the 1992 Loma Prieta earthquake in San Francisco, for instance, or in the extraordinary self-sacrifice of so many New Yorkers during the 9/11 terrorist attack – than in times of ease, when squabbling and petty competition seem to be the rule. Lately, historian Andrew Delbanco of Columbia University has reinforced not only the outlines of the original American Dream – he sees it as a focus on the transcendent (God, Nation) which manifests itself in practical, community-based cooperative action – but also just how far indeed twentieth-century America had strayed from those early ideals.[23]

Many of the cooperative traits of such self-sufficient villages were carried to their logical term by the famed Shaker communities: 'Hands to Work, Hearts to God.' The Shakers continued and in some aspects intensified the communal early American pattern, itself really a vestigial form of the medieval European village, although elsewhere such communalism was already giving way – either to the rough-and-ready pioneer mentality, or to the cities as commercial centers. Right through the nineteenth century and into the twentieth, each Shaker village was more or less self-sufficient in normal times; yet all participated in a larger pattern of shared faith, institutions, and techniques.

After he and Nathaniel Hawthorne visited a Shaker Village in the Nashua River Valley during a walking tour in the Autumn of 1842, Emerson was moved to compare the 'socialist' Shaker community to a single great 'capitalist.' (Before Marx, one might make such comparisons.) He wrote:

> *They have fifteen hundred acres here, a tract of woodland in Ashburnham, and a sheep pasture somewhere else, enough to supply the wants of the two hundred souls in this family. They are in many ways an interesting society, but at present have an additional importance as an experiment of socialism ... What improvement is made is made forever; this capitalist is old and never dies, his subsistence was long ago secured, and he has gone on now for long scores of years in adding easily compound interests to his stock. Moreover, this settlement is of great value in the heart of the country as a model-farm, in the absence of rural nobility ... Here are improvements invented, or adopted from other Shaker communities, which the neighboring farmers see and copy ...*[24]

Indeed, the Shakers never shared the bias against technology of the Amish or many another early American utopian experiment. For the Shakers, the Kingdom had already come. They considered their foundress, the redoubtable Ann Lee, to have been the Second Coming. If they were to live up to their vision, they had to set about creating the conditions of Paradise right here on Earth, in their own communities. Each settlement was to be an example to the rest of the world: "No one will find a spiritual heaven," they used to say, "until they first create an earthly heaven."[25]

What would a redeemed human life look like here and now? Coming from all walks of life, they knew the unsettling effects of unbridled sexuality, and voluntarily adopted a celibate lifestyle. Yet there were always children aplenty in the Shaker villages, and fine schools for them, since the Shakers started some of the first American orphanages. Knowing from experience the evils of greed, they agreed upon communal ownership of lands and goods. Having once felt the sting of persecution, they advocated religious toleration a full century before the word 'pluralism' became fashionable. And by undertaking collectively the day-to-day chores of maintaining their communities, they hit upon some remarkably practical improvements. They were among the first to package and market seeds, herbs and spices. And they invented, among other utilities, the flat broom, the circular saw, the washing machine and the clothespin: all labor-saving devices designed to free people from onerous, repetitive tasks.

Their simple, hand-tooled furniture is now by far the most valued body of American antiques. The Shaker way of life made tangible many of the ideals – thrift, efficiency, industry, inventiveness – that helped build the young American nation. The works they left behind share a certain aesthetic not only with Danish modern furniture, but, indeed, with many of Fuller's own artifacts: absolutely no frills, a functionalist minimalism which results in a spare and simple elegance of form. There is one very big difference. Shaker artifacts were all hand-made, and therefore each unique. Indeed, some scholars say that the Shaker way of life succumbed to industrialization and mass production, which eventually cut into the market for their products. And Fuller, more than Hart Crane or any other American, not only built with industrial tools but saw himself as the epic poet of industrialization. It's a refraction we shall examine a little later.

Although 'artificially' re-opened for newcomers in 1999, the original Shaker Covenant was intentionally closed in the late 1980s by the last few elderly Shaker Sisters, who have since passed away. The end for this living focus of the American dream came, says scholar Leonard Mendelssohn, neither from their celibacy nor from industrialization. It came, says he, because America is no longer capable of facing up to its own ideals, let alone sustaining them.[26]

Yet even in their passing, the Shakers passed on something of that dream and those ideals. It's still there; you feel it whenever you visit the quiet, simple buildings and manicured park-like grounds of one of their model settlements. Here there are no ruins, only the well-preserved shell of a former life, a tidy utopian vision frozen in time and in place, perhaps awaiting a new realization. Here you see the remnants of a vision that was more or less fully realized, and that lived out its natural life-span.

The neglected, burned-out, rubble-strewn shell of Bucky Fuller's Expo Dome, on the other hand, would strike the latter-day visitor in exactly the opposite way. Here, too, was a utopian vision, but of what? Where did such a powerful idea come from, and why has it all been left to fall apart? What was intended to happen here? And what would Fuller's vision look like, had it too been lived out and fully realized? Before even trying to answer, we need to excavate the American Dream just a bit further, in order to discern the peculiar way it informs and intersects Fuller's own vision.

The Original Vision

The American dream has always been to some extent an immigrant's vision – of economic opportunity, of political liberation, of back-to-the-Earth rusticity, or religious millenarianism. But through this lens, we see only what the newcomers *sought;* the other face of privations suffered in their homelands. Yet the American dream was also what the new arrivals *found* when they arrived on these American shores. Indeed, if we permit ourselves to dig down a little deeper in our speculative archaelogy, there seems to be a very special ingredient rooted in the native soil, a unique spirituality to this bio-region, a mindscape that also belongs to this landscape, part and parcel of the distinct seasons and mountains and changeful winds, and part also of the human character these all tend to mold.

In this sense, Emerson and Thoreau in their self-reliance, as well as the Shakers in their Kingdom-already-come, were themselves only inheritors of the original American dream. The early Americans were English or European men and women who found themselves in a rich and abundant land, which they took to be a gift from God. In this attitude, they were not alone. And from this angle, it may well be a very old dream indeed, from an ancient people who never acted without taking into account the effects of their every action down to the seventh generation.

Partly *because* Bucky rarely took note of cultural context, it falls to us to wrestle with the implications of the fact that he built his Expo Dome on Indian land. The Northeastern Woodlands sheltered native inhabitants of two distinct language families: the Algonkian speakers were mainly nomadic hunters; their less numerous Iroquoian neighbors were both hunters and agriculturalists, living in settled communities, and visiting the St. Lawrence and Montréal areas only as distant hunting grounds.

The Shakers, as it happened, first settled in the Mohawk River valley, near present-

Montréal Expo Dome (© Robert Duchesnay,1986.)

day Albany, NY. Only a few *months* before the Shakers arrived, this area had been at the heart of the Mohawk Nation, Elder Brothers of the Iroquois Confederacy. The first Shaker meeting houses even echoed the Mohawk longhouse architecturally, with its open dancing floor and separate entrances for men and for women. And when the Shakers danced there the ecstatic, shuffling 'Bruin' dances that became their namesake, they recorded that 'old Indian Spirits' sometimes spoke to them. What did the early Shakers hear in those old Indian voices?

We may only speculate, but surely they knew they were setting up their community on Mohawk lands. The Iroquois Confederacy embodied not only the first disarmament treaty ('burying the hatchet' under the white roots of the great Tree of Peace) but the first genuinely United Nations in human history (which in turn became the prototype for the division of powers in the eventual US Government). The Iroquois League would have been the immigrants' first encounter with a fully functional democratic society, coming as they all did from the old and despotic monarchies of Europe. If these 'savages' can do it, Benjamin Franklin once wrote, why can't we? He, Madison and Jefferson made sure that the Iroquois Chiefs were consulted in drafting the original US Articles of Confederation, modeled on their Confederacy, and the Iroquois were the first to recognize the new American nation even after it took a more centralized, federal turn.

The Shakers would have seen what was left of the fertile farmlands and orchards the native peoples had left behind when the newly federalized Americans turned upon them. This may indeed be the saddest part of our archaeology of the American dream buried in Bucky Fuller's work. In 1779, George Washington ordered General Sullivan and his 3,000 soldiers to totally destroy the villages of the Iroquois Confederacy. It was one of the most shameful episodes in American history – from our angle, no less than an attempted murder of the native American dream – and deserves a moment of our reflection. Historian Page Smith writes:

> *Sullivan's campaign was the most ruthless application of a scorched-earth policy in American history. It bears comparison with Sherman's march to the sea or the search-and-destroy missions of American soldiers in the Vietnam war. The Iroquois Confederacy was the most advanced Indian federation in the New World. It had made a territory that embraced the central quarter of New York State into an area of flourishing farms with well-cultivated fields and orchards and sturdy houses. Indeed, I believe it could be argued that the Iroqouis had carried cooperative agriculture far beyond anything that the white settlers had achieved. In little more than a month all of this had been wiped out, the work of several generations of loving attention to the soil ...*[27]

But the dream did not die. It metamorphosed, as we have seen. So what did it mean to begin with?

The interpreter of dreams may be allowed a certain latitude not permitted the historian ... What we glimpse from the very beginning is a dream or, if you like, an ideal of *cooperation*. Even the most 'rugged' individualist cannot stand alone for long through the harsh winters of the Northeast Atlantic region. The European model had always been one of *dependence* upon a central authority: *mon*-archy, one leader, one principle. The rhetoric of the emerging American republic was a cry for *independence* from all that, autonomy. But the reality of their lives puts the rhetoric in perspective. The picture we see is one of bold individualists indeed capable of standing alone in their thinking, but inevitably drawn into natural alliances with one another: the Shakers trading with the new American polity, giving and taking selectively; Emerson standing by the younger Thoreau in the latter's civil disobedience and rejection of society's constraints and conventions. Solzhenitsyn's hermit posture was not a native New England stance; and indeed, he could hardly have maintained his absolute privacy without the aid of his neighbors. Even at the very start, when John Winthrop and the Puritans came over the Atlantic in 1630 to found the Massachusetts Bay colony, their biblical 'city upon a hill,' they saw that the pilgrims from the Mayflower voyage nearly a decade earlier had already made common cause with the indigenous inhabitants. The only native American religious holiday, still celebrated by Americans and Canadians alike, purportedly commemorates the harvest they celebrated together so long ago. The Pequods of Massachusetts were eventually exterminated by the colonists in the War of

1637. The neighboring Mohawks, though literally decimated to one-tenth of their original numbers, did however survive with most of their customs, tribal structures, language, and lore intact.

> *Our creator made all of life with nothing lacking. All we humans are required to do is waste no life and be grateful daily to all life. And so we gather all our minds into one and send our greetings and our thanksgiving to our maker, our creator.*

So ends one version of the famous Thanksgiving Address of the Iroquois League, the proper ceremonial opening for any important occasion, which often takes the better part of an hour to perform.[28] The pilgrims didn't really 'originate' Thanksgiving at their Plymouth colony in December, 1621. They merely 'observed' in their own way the harvest festival and thanksgiving ceremony of the native inhabitants, along with the appropriate native foods – wild turkeys, pumpkins, cranberries – which enabled them to survive that first long, cold winter.

Mark the main themes: Life – not just as a given, but as *gift.* The clear mind is the thankful mind, say the Mohawk elders. And Nature – not just as raw material for 'development,' but as *the way things are.* Waste not, want not. Finally, interdependence – the rule of Life for people, indeed, but in reciprocity with the entire natural world. When Christians say grace, they thank God for the food. But the Indians always thanked the food, too. The traditional Thanksgiving Address scales the entire Creation – Earth, water, animals, trees, edible plants, birds, the sun, moon and stars – before addressing its final thanks to the Creator. When the Europeans departed for the 'New' World, some came by reason of faith alone, others put their faith in reason alone, but all had voluntarily cut themselves off from their roots in their own native soil. Yet once they arrived, they found a living network of vital relationships already awaiting them on 'New England' soil.

Long before this first meeting of Europeans and Native Americans, the peoples of this continent had known, and recognized, a great genius – a figure quite as remarkable in his own context as a Jesus of Nazareth, or a Gandhi, or a Martin Luther King in theirs. His own name was Deganawidah, but it is sacreligous to call him by name. He is remembered only as the Peacemaker. He was a Huron, who came unarmed to Mohawk territory (mainly encompassed by the central quarter of today's New York State). Together with his disciple Aionwahtha (Hiawatha), he began a process which eventually brought peace to five, and later six, tribes long at war with one another: Seneca, Onondaga, Cayuga, Mohawk, Oneida and, later, Tuscarora. Out of his vision, steadfast courage and remarkable political innovations, the Iroqouis Confederacy was born. Scholars and native peoples disagree as to whether this all happened in the century or two before the white settlers arrived, or several centuries earlier. But in any case, and contrary to the many Western preconceptions which Robert Vachon's extensive and intensive intercultural work convincingly debunks, the Confederacy was neither an empire nor a nation of warriors.[29] The Peacemaker outlined a set of *kinship* relations, a great extended family of peoples – Clan Mothers, Elder and Younger Brothers, etc., – each with different languages and customs, who nonetheless agreed to meet in peaceful council at a Central Fire, rather than imposing a single 'kingship' regime of the sort familiar to Europeans.

What the Peacemaker taught instead was **KAYANEREKOWA**, the Great Law of

Peace, and the people who lived according to his teachings came to call themselves the People of the Great Peace. They still do. Even today, in the face of all sorts of political, military, religious, and social pressures from the modern American and Canadian nation-states, the Confederacy lives on. And here, I submit, we finally touch ground. Here may well be the oldest stratum, the foundation level of the American dream we've been trying to excavate.

During the final decade of the twentieth century, the Mohawk communities straddling the borderlands between the United States and Canada were torn apart by bitter conflicts, both internal and external. So-called Warriors supported by gambling money and cigarette smuggling were vying for power with the Band Council, supported by fat paychecks from the Canadian government. An armed stand-off with the Canadian army in 1991 went on for weeks. The traditional people, still the majority, were caught in the middle. Weary of all the fighting, and frustrated that the Iroquois Great Law of Peace has been so blatantly flouted, some of them determined to return to their ancient homeland, the 'Land of Flint' on the Mohawk River near Albany, NY. It is called the 'Clean Pot' movement, implying a fresh start. Some kind soul has deeded a parcel of their original land back to them, so that such a move at last looks feasible, although the politics and logistics may yet turn out to be a nightmare.[30] Iroquois people have always been a dream culture, following their dreams even if they sometimes seem to defy the logic of the daytime world. Once upon a time, you recall, they built a dome in Montréal for Bucky Fuller. And today some of the elders are dreaming of a return to their homeland and traditional ways, a visionary feat which would have to span the highest idealism with the most down-to-earth practicality. Once again, they are building without scaffolding. Maybe there will always be American dreamers, even if America at the turn of a new millennium remains oblivous to them.

The Way Things Are

By now you may well be wondering what all this ancient history has to do with Bucky Fuller and his domes. Patience; we are nearly done digging. There is more at stake here than collective amnesia about the original American dream, or the bald fact that it was the Mohawks who actually built Bucky Fuller's Expo Dome in 1967 on Ile St. Hélène, itself once Mohawk land. At the deepest level, what is at stake is a matter of principle. Indeed, the Iroquois Great Law calls into question the very way we conceive of 'laws' or 'principles' of Nature.

Westerners tend to think of 'law' as something man-made, a principle or set of principles discovered by the human mind and imposed upon the things of the natural world or, indeed, upon human nature itself. The Iroquois Great Law, by contrast, is a recognition of the Law *already there* in Nature herself, an attunement to the 'Way' things are – the original plan of the Creator. Westerners, moreover, tend to see the natural world as fundamentally flawed, or incomplete, or even sinful; it is merely raw material to be developed to human ends. The Iroquois, in concert with many traditional societies the world over, have always seen Nature as pretty good and complete just as it stands: an order, a peace and harmony to be preserved.

This, then, is the first Law, to which human life, customs and institutions ought to be attuned. It is a way of Life already 'written' into the patterned energies and dynamisms

A Tale of Two Museums

The American dream of 'cooperation' between different peoples leading to a new polity, always in tune with the reciprocating 'way of Nature,' has been articulated in many ways. Largely, the holistic vision we have discerned at its root has been displaced by monistic and dualistic models – the ideal of cooperative 'confederacy' displaced in America by compulsory federalism with its 'melting pot' ideal, and in Canada by competitive 'provincialism' with its own ideal of a cultural 'mosaic.' (Which may be why each polity is so susceptible to its contrary: bland conformity in Canada, cut-throat competition in America.) And just where might we expect to find these two visions, the mosaic and the melting pot, enshrined and architectually embodied? In the 'national' museums, of course ...

The Smithsonian in Washington, DC: a hodge-podge of buildings designed in different epochs for different purposes. A place for everything, everything in its place. 'Civilization' resides in the artifacts from mainstream Western sources, duly preserved in the arts museums, as well as in the aerospace museum where even the 'future' is enshrined. All other cultures, even native American until very recently, have been lumped together in the 'Natural' History museum next to the stuffed animals and dinosaur skeletons. One big melting pot.

Canada's Museum of Civilization in Ottawa: one design concept, by a single architect – flowing tiers of stone set into the curving river, the flow of time as organizing principle. 'Civilization' here is the upsweep of evolution from the Grand Hall of native artifacts through the many waves of immigrants – Basque whalers, French settlers, English soldiers – all the way up to a nineteenth century Main Street. You follow the pattern by climbing the stairs, but nobody's allowed to peek into the attic where the future must be hidden. Unfinished mosaic.

Surprisingly, something else has recently surfaced, within yet beyond both images: Bill Reid's huge Black Canoe jam-packed with shape-shifting spirits from Haida Gwaii today faces the Smithsonian from the Canadian Chancery, while its white plaster cast now adorns the Museum of Civilization. The feat of bilocation is telling. Each mythic form – Bear, Wolf, Eagle, Beaver, etc. – strains to pull the boat in its own direction, yet each is pulled inescapably into the medley of other forms. Raven is at the helm and his way, 'so I have heard,' is to go through endless transformations.

Maybe it's time we North Americans found a new metaphor for ourselves: whether figured as a mosaic or a melting pot, it's beginning to look like we're all in the same boat.

of the natural world: all things 'are' always and only in intimate and constitutive *relation* with one another, i.e., *interdependence.* All that is needful for a fully human life has already been given. The way to realize this law in human affairs, it follows, is neither by compulsion nor competition, but by cooperation – with one another, with Nature herself, and with the spiritual power, *manitou,* surging through her every fiber.

The Shakers no doubt felt and lived it too, in their millennial Christian way. For them, as for most medieval and renaissance thinkers, the Book of God's Word (the Bible) and the Book of God's Works (Nature), were ultimately the same Book. "I never feel closer to God," said Eldress Gertrude Soule about a month before she died in 1988, "than when I am walking in these forests and fields and pastures." Emerson and Thoreau, too, borrowed a leaf from that Book when they spoke of communion with the very 'Soul' of Nature. In 'Worship,' a late essay from his *Conduct of Life,* Emerson took to task any soft-headed moralists who ignored this connection between Spirit and Nature:

> *The true meaning of spiritual is* real; *that law which executes itself, which works without means, and which cannot be conceived as not existing. I find the omnipresence and the almightiness in the reaction of every atom in Nature. I can best indicate by examples those reactions by which every part of Nature replies to the purpose of the actor, – benificently to the good, penally to the bad. Let us replace sentimentalism by realism, and dare to uncover those simple and terrible laws which, be they seen or unseen, pervade and govern.*[31]

And for Buckminster Fuller, the single word which sums much of this up is *synergy.* It is a Greek word, which basically means 'working together.' Now that every other business in the Western world seems to have adopted 'synergy' as its middle name – a tendency which hit a new low when Mobil Oil trade-marked its latest multi-additive gasoline concoction *Synergy*® – it may be worthwhile to glance back at the word's venerable origins. It is a theological word which retained currency in Fuller's time solely in chemistry, for the unique behaviors of compounds. Yet St. Paul had used it in his Epistles (Rom. 8:28; I Cor. 3:9) to illustrate not a static but a dynamic conception of human, divine and cosmic cooperation:

> *I did the planting, Apollos the watering, but God made things grow … We are fellow workers (*synergoi*)*[32] *with God; you are God's farm, God's building.*

In his modern and scientific context, Fuller uses the word 'synergy' to describe "the behavior of whole systems unpredictable from the behavior of their parts taken separately."[33] He liked to claim it is the only word in English which describes this unique behavior of whole systems, always something greater than the sum of their component parts. By his time – after the mechanized slaughter of the First World War and the killer flu that followed in its wake – religion and politics no longer seemed to be viable arenas for utopian visionaries. They had pretty much yielded the field to science and technology. Fuller, very much a man of his historical moment in this respect, considered all political and religious systems parochial and atavistic. Yet he was their inheritor. He inherited above all the belief in open-ended progress – itself part of the age-old Western quest to transcend this world – in its streamlined, evolutionary twentieth-century form. Like Francis Bacon, he put his faith in the empirical methods of science. Following the utilitarians, he sought the greatest good for the greatest number of his fellow humans ... though he did not find this 'good' in terms of the greatest pleasure (Bentham, Mill), nor the greatest financial gain (Adam Smith, Ricardo, Keynes), like so many of his contemporaries.

And he made a great discovery. He was not the first to see it, to be sure, but I would say he was the first to see clearly these metaphysical patterns which govern physical changes through the hi-tech lenses provided by modern science. How did he find it? The novelty of his approach was the very personal way he took Einstein's relativity as a license to re-envision the entire universe of human experience. Instead of a static Newtonian world of separate 'things' whose normal state was to stay pretty much the way they always were, Fuller saw the universe as a dynamic complex of interconnected energy events: a *fluxus quo.* Can we discern *the form the flux takes?*

In mapping the structures of this dynamic universe, Bucky deliberately chose, moreover, not to follow the piecemeal, hyperspecialized approach of most science in his day. Fuller's understanding of himself as a scientist took its inspiration from Sir Arthur Eddington's definition of science as the systematic attempt to set in order the facts of experience. In this sense, despite his very up-to-date scientific information, he was more

like a natural philosopher in the nineteenth-century English (Lyell, Darwin) or American (Emerson, William James) mold, still seeking to embrace the Whole. Indeed, one is reminded of Galileo's early formulation of the scientist's quest to 'read' the Book of Nature in a comprehensive way:

> *Philosophy is written in that very large book that is continually opened before our eyes (I mean the universe), but which is not understood unless one first studies the language and knows the characters in which it is written. The language of that book is mathematical and the characters are triangles, circles, and other geometric figures.*[34]

In a sense, Buckminster Fuller claimed nothing less than to have read that Book, in its original language of images. *Synergetics* is his exegesis of it, a *grammar* of this language of images. What Bucky did uncover – or 'discover,' if you suppose as he did that he was the first to do so – was a matrix of interconnections which also turned out to be the matrix for all these transformations: "Nature's Coordinate System," he boldly called it. Here is Fuller recounting his own quest in *Utopia or Oblivion:*

> *In 1917, I found myself asserting that I didn't think nature had a department of chemistry, a department of mathematics, a department of physics, and a department of biology and had to have meetings of department heads in order to decide what to do when you drop your stone in the water. Universe, i.e. nature, obviously knows just what to do, and everything seemed beautifully coordinate. The lily pads did just what they should do, and the fish did just what they should do. Everything went sublimely, smoothly. So I thought that nature probably had one coordinate system and probably one most economical arithmetical and geometric system with which to interaccount all transactions and transformations. And I thought also that it was preposterous when I was told that real models are not employed in advanced science, because science was able to deal with nature by use of completely unmodelable mathematical abstractions. I could not credit that universe suddenly went abstract at some micro-level of investigation, wherefore you had to deal entirely with abstract-formula, unmodelable mathematics ... I thought then that if we could find nature's own coordinate system we would understand the models and would be able to develop much higher exploratory and application capability. I felt that if we ever found nature's coordinate system, it would be very simple and always rational.*[35]

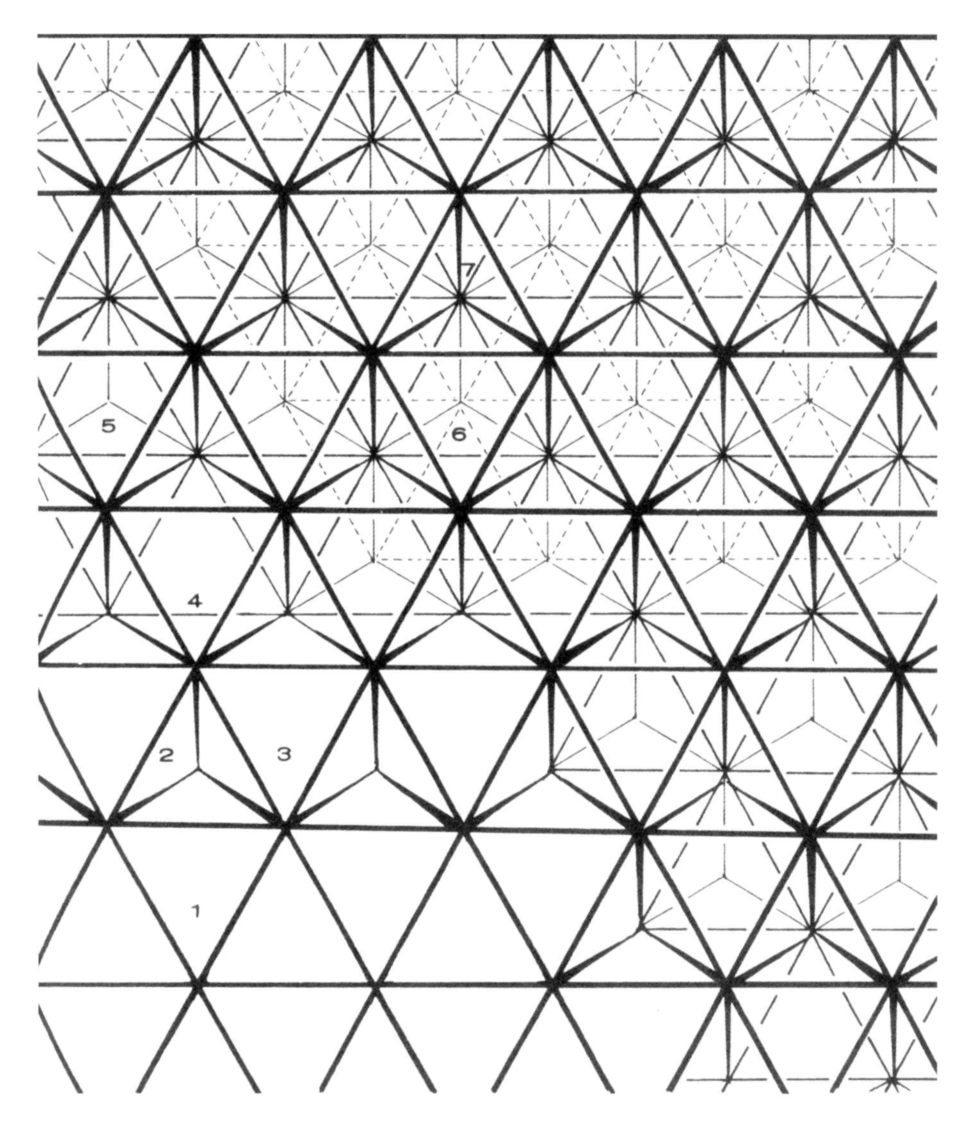

The Coordinate System of Nature, alias Octet Truss. Thickest lines are elements nearest you and (despite effect of perspective) all members are the same length. Start at bottom left. 1. First-level triangular grid. 2. Tetrahedra, pointing away from you, with octahedra (3) appearing between them. 4. Second-level grid joins tips of tetrahedra. 5. Next array of tetrahedra. 6. Dotted lines show third-level grid. This can be continued indefinitely. Twelve-way vertices (7) in second level are centers of vector equilibria.
(From Hugh Kenner, Bucky – A Guided Tour of Buckminster Fuller, *NY (Morrow) 1973.*
By permission of the William Hugh Kenner Estate.)

Montréal Expo Dome, 1967 (Estate of R. Buckminster Fuller.)

Geoscope

Bucky Fuller wanted to build a Geoscope under his Montréal dome. His original plan was rejected by the US Information Agency, the ever-smiling propaganda arm of the US government, in favor of a display featuring gee-whiz scientistic gew-gaws like a monorail ride and a space capsule, American pop culture pin-ups like Marilyn Monroe, and some picturesque mylar balloons and multi-coloured parachutes floating in the dome's upper reaches. So for reasons of its own, the US government gutted Fuller's dome long before fire finished the job. He was retained as architect of the building, which amounted to little more than the shell of his original vision for Expo '67.

But what, you may ask, is a Geoscope?

I have been at pains to bring the New England context of Fuller's vision up into high relief in order to emphasize its distinctively American character. Although global and in some ways abstract, both what Fuller was looking for and what he found had local roots. Indeed much of the world today sees 'globalization' as nothing more nor less than global 'Americanization,' although Americans tend not to see the baggage they carry with them into the world at large. This paradoxical contrast – global vision as a local phenomenon – is all the more striking since Fuller himself so relentlessly emphasizes its planetary consequences. His Geoscope was an idea, first hinted at in his early *4D Time Lock*, and never in fact fully realized. It was to be *a miniature Earth* through which one could discern all sorts of trends and phenomena not detectable by ordinary human senses – movements of populations, resource distribution, long-term patterns of discovery (e.g., of the periodic table) or invention. It would permit computer modelling on a vast scale, and serve as the ultimate playing field for the World Game, Bucky's inventory of Earth's resources and deployments. It was one of the main tools he hoped to use to 'make the world work.'

What then is the connection between the local roots I have been stressing and Bucky's global, indeed planetary, vision? There are many factors, of course – extrapolations of his 'geometry of thinking,' for example, as we shall see later – but I would single out one: *the sea.* New Englanders have always been seafarers, and Bucky was no exception. Fuller looked at the planet with a sailor's eye; he always claimed that his generalized Navy training as a midshipman in the First World War was his introduction to comprehensive, world-around thinking. Seafaring technology inspired him by its elegant efficiency, and the seafarer's world-around experience lent credibility to his own planetary 'designs.' Landlubbers take 'solid' things to be real; they find comfort in national sovereignties, in the assurances of institutionalized religion, in the reforms advocated by politicians. Fuller knew that once you set sail, it's just you and the elements. You leave your drawing room theories behind, and learn to survive in the real world. You have to know how to sail with the wind, and when to tack into the wind. You have to work with Nature at her most elemental if you're going to get where you want to go.

Now the Earth's land masses are all ultimately islands in the one great interconnecting World Sea, just as the 'solids' or compressive members in Fuller's tensegrity figures are 'islanded' in an elastic continuum of tension. Go back to Herman Melville's exotic South Seas journeys, or recall the uncannily precise knowledge of the world's oceans his New Bedford whaling men possessed in *Moby Dick,* and you see that New Englanders who

Cornell University Geoscope (Estate of R. Buckminster Fuller.)

made their living from the sea were thinking in world-around terms for a couple of centuries before Bucky Fuller. The sea was Fuller's unfailing source of inspiration. Whether spinning his tales of Polynesian genesis – where humans evolved in the warm waters of lagoons in the South Seas (an hypothesis contending quite dramatically over the past decade with the conventional notion of an African genesis); or his cautionary tales about the Great Pirates – who knew and exploited the sea routes while everybody else squabbled over chunks of dirt; or deriving his geometry from watching bubbles form in the wake of a ship – "I knew then that Nature doesn't use *pi*"; Fuller returned to the sea in his 'cosmic fishing' for ideas at least as often as he returned to his family's Bear Island estate off the coast of Maine.[36] He had houses in many places – a dome in Carbondale, Illinois, an office in Philadelphia, another in Los Angeles – but he always seemed most at home at the helm of a sailing ship. When he finally acquired a seventeen-ton, ocean-going sloop of his own, he christened her *Intuition*, and as a memorial of the occasion wrote an epic poem of the same name which remains one of the most concise and readable summaries of his thinking.

For the landlubber, the sea divides. For the seafarer, the sea connects. The difference is crucial for Fuller's thinking, and for the way he built things. What most of us see as solid – the bricks, for example – Fuller sees as discontinuous and often redundant. What most of us see as discontinuities – the mortar, to stick with our example – Fuller sees as a continuum, the tensional structure holding any ordinary bricks-and-mortar building together. Hugh Kenner, in his *Bucky, A Guided Tour,* cites many other instances: When most people look at plywood, they see the wood. Bucky saw the glue.[37] And, even more tellingly, when most people look at planet Earth, they see over 180 nation-states, each with clear and distinct borders. Fuller saw one island Earth in one World Ocean, not unlike the view soon popularized by photos from the NASA space flights.

The flat Earth mentality Fuller railed against is a landlocked view; order is to be found only within lines we ourselves have drawn. Landlubbers think the sun rises and sets upon their own fixed position. On the ocean, however, one senses the swell of the waves, the ebb and flow of the tides, and even the curvature of the Earth. At night, your sail slices between the sea and the Milky Way. When Fuller called the Earth a space ship, people criticized him for comparing the planet to one of NASA's tin cans floating in the void. Such a view would indeed be severely reductionistic. Yet Fuller's perspective was exactly the reverse; the Earth is the proper model, a fully-provisioned sailing craft winging amongst the stars: we are all astronauts.

In the ambitious lecture given in 1962 and published as *Education Automation,* Fuller already foresaw an electronically 'wired' world gathering all its fundamental survival information to be fed into a Geoscope 'hub' to help coordinate 'total world planning' and facilitate educational 'comprehensivity.' He would no doubt be dismayed, now that such a Worldwide Web is up and running, to discover how today's Internet has veered into the further fragmentation of purely self-interested groups furthering their own ideological agendas, rather than congealing into the total learning 'environment,' served by access to all the world's electronic media, that he had in mind when proposing the Geoscope in the 1960s.

It was probably no coincidence that Fuller finally became known to the world in the mid-1960s, at the time when photos of Earth from space began to transform conventional outlooks and even to transfer many people's allegiances from political or religious sovereignties to the planet as a whole. The mythologist Joseph Campbell saw in those photos the beginnings of a planetary mythology, encompassing not only a new environmental sensitivity, but a new spirituality as well. Of course Fuller had seen it all already, in his own way. The first practical application of his synergetic/energetic geometry, as it happens, was the Dymaxion Air/Ocean Map, using an icosahedral projection to divide the Earth's surface into twenty triangles, which could then be cut out and laid flat in such a way that either the land masses were continuous, and surrounded by water, or the one ocean was continuous, and surrounded by land. Amazingly, his was the first, and so far it is the only, flat cartographic projection in human history which does not distort the actual shape of the land masses. Early in the 1990s, *National Geographic* magazine commissioned a 'new' world map and, ignoring Fuller completely, produced yet another variant of the Mercator projection which not only still distorted everything at the poles, but again centered on the North Atlantic, a purely political (one might even say colonialistic) orientation.

Now besides the realistic shapes, such a map is also useful for plotting quantities – of things, people, trees, cars, telephones, televisions, etc. – in a way that makes world-around trends often invisible (the deforesting of the temperate zones over the past couple of centuries, for example, or today's rapid destruction of the rainforests) immediately evident to anyone with eyes. Fuller's Geoscope idea derived directly from the Dymaxion map; it was the 3D version, and since trends over time could be modeled with a little help from computers, it was potentially a four-dimensional, 'virtual' Earth. We've all been to planetariums for simulated trips to the moon or the planets. Bucky wanted to build a unique kind of 'Whole Earth' planetarium, if you will, which would also serve as a comprehensive database of all human knowledge of the Earth and her resources – precisely the sort of compendium first proposed by Bacon 400 years earlier in his *New Organon*. No doubt Bucky drastically underestimated the quantities of data involved, just as he over-estimated the computing capability of his time. Even Japan's Earth Simulator, the latest evolution of Fuller's idea and the largest supercomputer currently up and running, attempts only to handle weather simulations, not the entire past history as well as a gamut of possible future scenarios for 'Spaceship Earth.' But if credit for the original idea of such a vast compilation of Earth-data must go to Bacon, credit for reviving the idea belongs to Fuller, with perhaps a nod of appreciation to the folks who took him up on it in the early *Whole Earth Catalogs*. He reviewed the concept retrospectively in *Critical Path:*

> *I told the United States Information Agency in 1964 that by 1967 the regard of the rest of the world for the United States would be at its lowest ebb in many decades – if not in the total two centuries of the U.S.A.'s existence. Since each country's World's Fair exhibit would be well published all around Earth, I felt that it would be very important that the United States do something that would tend to regain the spontaneous admiration and confidence of the whole world. This could be done by inaugurating at Expo '67 a computerized exploration for the most universally creative and economically sound internal and external U.S.A. policy formulation ...*
>
> *On the working assumption that humanity had established implicit confidence in ... computers and automated instrumentation, I proposed in 1964 that the United States Expo '67 exhibition should have a 400-foot-diameter 5/8 sphere building similar in shape to the*

250-foot-diameter building actually built for Expo '67. In the basement of this building would be housed an extraordinary computer facility. On entering the building by thirty-six external ramps and escalators leading in at every ten degrees of circumferential direction, the visitors would arrive upon a great balcony reaching completely around the building's interior quarter-mile perimeter. The visitors would see an excitingly detailed 100-foot-diameter world globe suspended high within the 400-foot-diameter 5/8 sphere main building. Cities such as New York, London, Tokyo, and Los Angeles would appear as flattened-out, basketball-sized blotches with the tallest buildings and radio towers only about one-sixteenth of an inch high.

Periodically the great spherical Earth would be seen to be transforming slowly into an icosahedron – a polyhedron with twenty (equilateral) triangular facets ... Slowly the 100-foot-diameter icosahedronal Earth's surface would be seen to be parting along some of its triangular edges, as the whole surface slowly opens mechanically as a orange's skin or an animal's skin might be peeled carefully in one piece ... The icosahedronal Earth's shell thus would be seen to gradually flatten out and be lowered to the floor of the building. The visitors would realize that they were now looking at the whole of the Earth's surface simultaneously without any visible distortion of the relative size and shape of the land and sea masses having occurred during the transformation from sphere to the flattened-out condition we call a map. My cartographic projection of the 'Sky-Ocean World' functions in just such a manner as I have now described.

This stretched-out, football-field-sized world map would disclose the continents arrayed as one world-island in one world-ocean with no breaks in the continental contours. Its scale would be 1/500,000th of reality ... When completely installed and ready for use, [the Geoscope is] oriented so that [its] polar axes are always parallel to the real Earth's north-south polar axis, with the latitude and longitude of the installed Geoscope's zenith point always corresponding exactly with the latitude and longitude of the critically located point on our real planet Earth at which the Geoscope is installed. As a consequence of the polar axis and zenith correspondences of the Geoscope mini-Earth and the real Earth, it will be found that the miniature Earth Geoscope's real omnidirectional celestial-theater orientation always corresponds exactly with the real omnidirectional celestial-theater of the real planet Earth.

Since the two spheres (mini-Earth and real Earth) are rigidly coupled together tangentially ... the geographical-geometrical orientation attitudes brought about by their respective axial rotations and orbital travel around the Sun will be identical. The Geoscope has the same relationship to the Earth as has one of the relatively small lifeboats mounted fore and aft on the davits of an ocean cruise ship to the big ship herself. If the big ship changes its course from north to east, the lifeboat does likewise. If the bow rises and falls in a head-on sea, so too does the bow of the davits-mounted lifeboat ...

The great map would be wired throughout so that minibulbs closely installed all over its surface would be lighted by computer at appropriate points to show various, accurately positioned, proportional data regarding world conditions, events, and resources. World events would occur and transform on this live world map's ever-evoluting face. If we had 100,000 light bulbs for instance, each mini-light-bulb could represent 40,000 people ... The bulbs could be computer-distributed to represent the exact geographical distribution positioning of the people. Military movements of a million troops would be dramatically visible. The position of every airplane in the sky and every ship on the world ocean would be computer-control displayed. Weekend and holiday exoduses from cities into the country

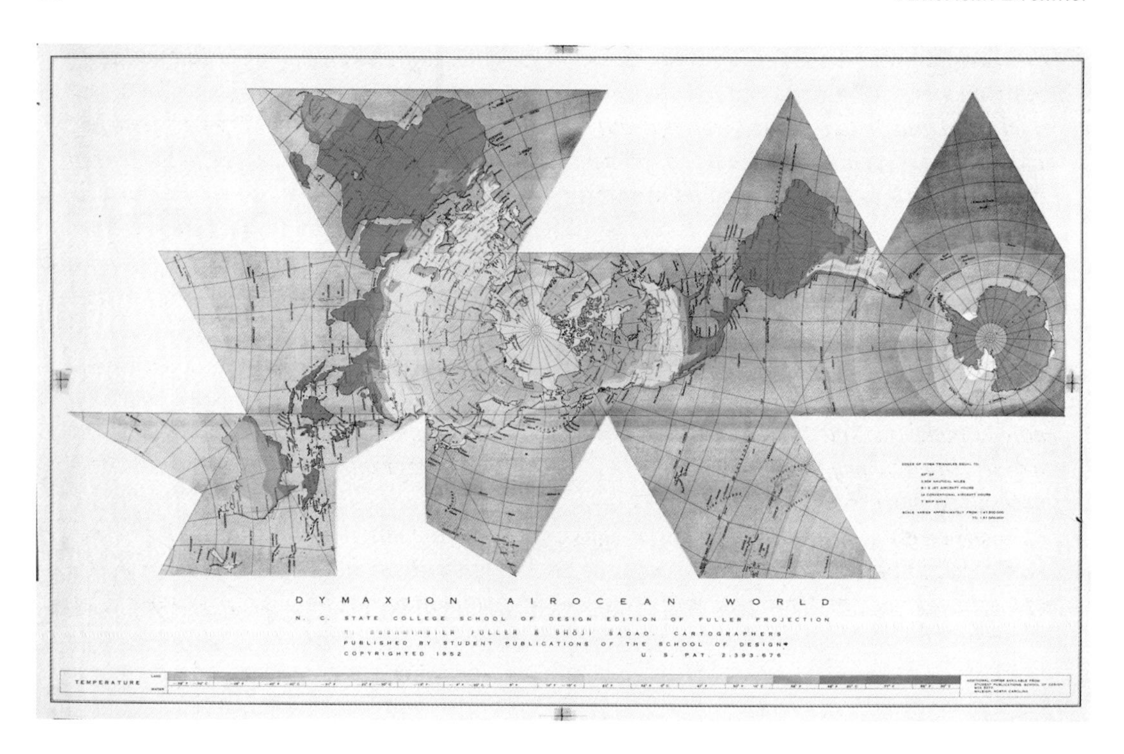

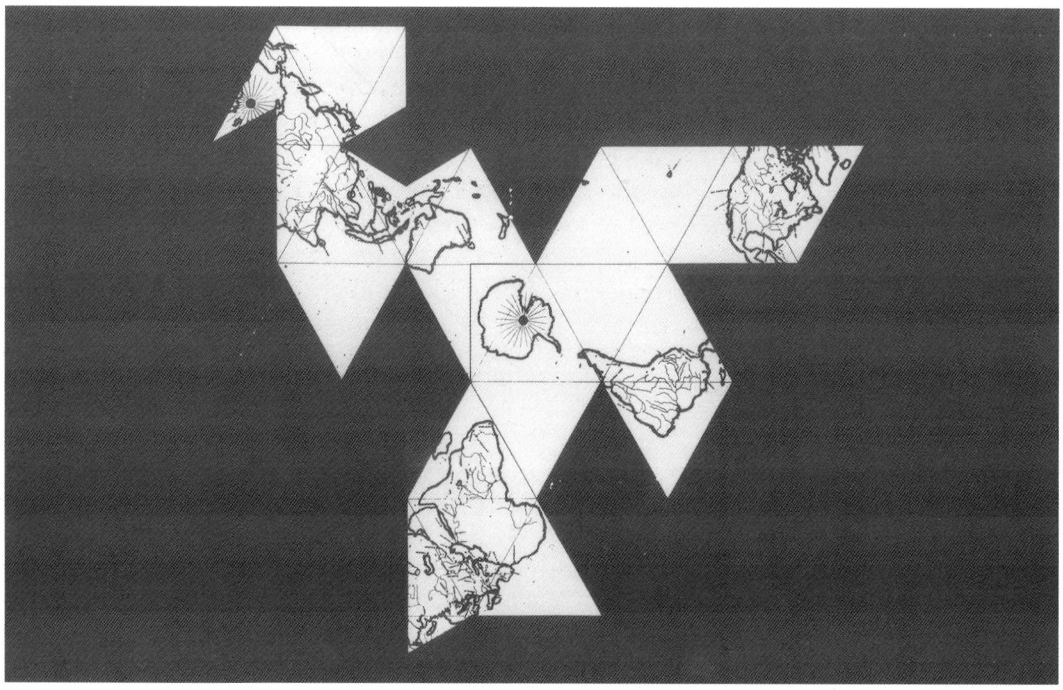

Top: Dymaxion Air/Ocean Map (1954).
Bottom: Dymaxion Projections re-assembled at South Pole to show British Empire's 'One Ocean World'
(R. B. Fuller and S. Sadao. Estate of R. Buckminster Fuller.)

or travel to other cities would be vividly displayed by computer-controlled tallying instruments.

I proposed that on this stretched-out, reliably accurate, world map of our Spaceship Earth a great world logistics game be played by introducing into the computers all the known inventory and whereabouts of the various metaphysical and physical resources of the Earth. (This inventory, which took forty years to develop, is now housed at my headquarters).

We would then enter into the computer all the inventory of human trends, known needs, and fundamental behavior characteristics.

I proposed that individuals and teams would undertake to play the World Game with those resources, behaviors, trends, vital needs, developmental desirables, and regenerative inspirations. The players as individuals or teams would each develop their own theory of how to make the total world work successfully for all of humanity. Each individual or team would play a theory through to the end of a predeclared program. It could be played with or without competitors.

The objective of the game would be to explore ways to make it possible for anybody and everybody in the human family to enjoy the total Earth without any human interfering with any other human and without any human gaining advantage at the expense of another.[38]

Synergy & Entropy

The Geoscope was nothing less than Bucky Fuller's vision for planet Earth in a nutshell. But of course it was never built, never even seriously considered by the USIA – although his proposal for a much larger Geoscope to be anchored in the East River outside the United Nations building in New York City did receive approval – if not funding – from the U.N.

What happened? Maybe certain powers didn't want "military movements of a million troops" in Vietnam to be "dramatically visible," let alone the deployment of many thousands of nuclear warheads. Indeed, with twenty-first century satellite imaging and remote sensing power, who knows what 'We the people' might see were we ever permitted a close-up look at the planet we live on? With fiber optic cable and broadband information superhighways already available, a big, centralized Geoscope like the one Bucky envisaged for Montréal would be only one possible display screen, perhaps for special effects like today's giant IMAX cinemas. Size, as Fuller never tired of pointing out, is not a generalized principle. He anticipated today's rapid miniaturization of everything with his concept of ephemeralization – 'doing more and more with less and less' – as the driving force in technological development. Indeed, there is no *technical* reason why everybody's home computer, which currently displays planetary information from weather satellites and the like, could not have access to long-range historical scenarios and future projections as well. But maybe most of us are content to accept our evening news from government, military and corporate media sources. Maybe we're just not as curious about the Big Picture as Bucky was, or supposed us to be. At any event, Bucky's Geoscope was never built. Something went wrong, something was out of synch …

Bucky would have calibrated his mini-Earth so that it would be oriented in the "celestial-theater" on exactly the same axis as the big Earth Mother Ship but, plainly, Fuller's dream was no longer America's dream.

I suspect that the rejection of his ambitious design for the Expo Dome did not come as a surprise to Fuller, who always thought big and knew that only portions of his projects would be realized in his lifetime. He had, moreover, worked for *Fortune* magazine in the 1940s, and over the years hobnobbed with the likes of Henry Kaiser and Henry Ford II. In later life Bucky claimed that in order for the 'Great Pirates' to pursue their multinational machinations unperturbed, the rest of us were being kept ignorant, preoccupied by the details of our own short-term getting and spending, and utterly unaware of any Big Picture. I had the chance to ask him, only a month before his death in 1983, whether there really was such a global conspiracy, or whether the sorry state of the world was simply the result of ignorance compounded over generations. "Young man," he replied, "I've lived on this planet 87 years. I've met these people, I know how they operate. And you better believe it's a conspiracy." I'm still not entirely comfortable with conspiracy theories (maybe the conspirators are ignorant, too, of what really will insure their own well-being?), but I can testify that the shock of recognizing the absolute degradation of the American dream was not so much Fuller's as it was that of an entire young generation who once believed that if you found a better way to live, or to do something, it would be welcomed and supported by the society around them. Not a chance … just ask some of the folks who have come up with improvements in automobile manufacture what Detroit has done with their ideas. In Vietnam, young Americans learned they were the bad guys. On the home front, they soon discovered they were the enemy. And when the system forces you to fight merely to save your own skin, the Big Picture does tend to fade into the background.

In the description of the Geoscope above, you might have noticed that Fuller draws an analogy between his project and a lifeboat. He was well aware that the big ship is sinking, at least if present trends prevail. Typically unconcerned about aesthetics, however, he did not remark that the "ever-evoluting face" of such a globe, mirroring the Earth's constantly changing contours with the resolution of a high definition television screen, might quite possibly have been the most *beautiful* man-made object on the planet.

Yet during those heady years of the 1960s and 1970s, there were still plenty of young people who wanted to play Bucky's World Game rather than war games – with or without a Geoscope in Montréal. After all, the Leakeys, reviewing the evidence on human origins they found at Olduvai gorge, had put forward the thesis that we humans became who we are because we learned how to cooperate and share food in interdependent communities. The dream of Emerson and the Shakers, not to mention the political innovations of the Iroquois Confederacy, now had a firm anthropological basis and, seemingly, a global arena.

Perhaps we have to step back to see such dreams clearly. Since the rise of *laissez faire* capitalism and the so-called Manchester School of economics with its 'free' market and 'invisible hand' in the early decades of the nineteenth century, there have been a range of vigorous alternatives to it – some religious, some secular, some vaguely socialist, some stringently Marxist, some tribal, some familial, some simply neighborly and communitarian. Taken together these trace a coherent alternative stream of thought about the foundations of human community. Bucky Fuller's thinking sails along with this current, even today driven underground by the prevailing mercantile propaganda. The New England Transcendentalists would be an early American example, interacting with, critiquing and reflecting on, various other such American experiments in community like Brook Farm or the Shakers. John Dewey's pragmatism, with its emphasis on consensus, sociability and cooperation rather than competitive individualism, would be another example of the continuing vitality of this alternative American dream. In England, John Ruskin's "law of help" would be the high water mark of the mid-nineteenth century, his response to what he called "the law of force," and the "illth" (assuredly not wealth) of the marketplace.[39] The same stream of thought enters the twentieth-century well defended by Pietr Kropotkin's *Mutual Aid,* a counterblast to Thomas Huxley's attempt to assert the imperial English designs and dominion of Social Darwinist values in the global marketplace of his day.

Half a century later, we find Bucky attempting to re-order the world into a series of interlocking service industries which would "take care of everybody" at a very high, technologically-enhanced standard of living. The Geoscope was to be a step in this direction: you have to know where in the world things are, and who needs precisely which things, in order to get them from here to there. In the 1980s and 1990s, Medard Gabel and others from Bucky's Philadelphia office were quick to put Fuller's inventory of "human trends, known needs, and fundamental behavior characteristics" onto a Hypercard program, in which form it was made available as "Global Recall" and "Global Data Manager" (you have my permission to cringe at the latter name). They were presumably anticipating the sort of decentralized home computer version of the Geoscope mentioned earlier, although they seemed to be a good deal more thorough at inventorying the physical resources than the metaphysical. Stewart Brand and his San Francisco Bay Area *Whole Earth Catalog* cohort of hip capitalists also got going with these ideas of Fuller's in the late 1960s, providing mail order 'access to tools' for anybody, anywhere, to build an alternative lifestyle to their own design – and they did include at least as much philosophy as physics. But by the time it was published by Random House as a gigantic bestseller, I suspect the final *Whole Earth Catalog* (1985) was less often used as intended, and more often as a kind of icon that sat unopened upon coffee tables or bookshelves. You had to have it, but you didn't really have to do anything with it. Nowadays that entire operation goes on as the *Whole Earth Review,* and is available in the software world for electronic cottages everywhere as WELL (Whole Earth 'Lectronic Link), a pay site that is now part of *salon.com* (major credit cards accepted).

As a matter of fact, I first encountered Fuller as a kind of intellectual curiosity in the pages of Norman Cousins' *Saturday Review* during the mid-1960s, but did not really catch his drift until the late 1960s, when everything seemed to be drifting his way. At the University of California, Santa Cruz, where I spent my undergraduate years, Fuller was one of the elements in a complex and exhilirating mix of new ideas with which we were going to save the world.

If Berkeley was political and controversial, Santa Cruz in the late 1960s was much quieter, more mystical and mythopoetic. There were about three million redwood trees on the often fog-shrouded campus, which helped give the place a mossy, timeless air. The university itself was a 'devolution' of the vast UC multiversity into something resembling the Oxford system of small colleges with the resources of a large university, where you got to know renowned professors on a first-name basis. Fuller's domes were going up on one side of my college – that would solve the world's housing shortage, eh? – while up the hill, thanks to the good offices of Paul Lee (a.k.a. the Herbal Trading

Company), Alan Chadwick and his disciple Jon Jeavons were hard at work putting in the first French Intensive garden to demonstrate the biodynamic agriculture that would feed everybody. "Our job is to MAGNIFY life," boomed Chadwick in his stentorian English stage voice, and he would plant by the light of the full moon. (Nowadays his method is hailed by the UN as the most effective, and least intrusive, kind of food production 'development.')[40] There were impromptu seminars on Fuller every week, led by avid students who later formed the Cathedralite build-your-own-dome kit company.

While I'm on this personal note, I might add something about the festive atmosphere – just in case you missed that party. There was always music in the air in those days, from visiting musicians like Ali Akbar Khan or Joan Baez or Keith Jarrett, who had just gone back to his acoustic piano – "Electricity is in everything" – or from students with talents not in the least hidden by their long hair and beards, or simply Jefferson Airplane or good ol' Grateful Dead on somebody's stereo. What was going on in Northern California at the time may well have been the last renaissance of the original American dream, although perhaps only historian Page Smith (heading up Cowell College) saw that aspect clearly through all the psychedelic pageantry. Norman O. Brown might be found by night leading Dionysiac revels through the redwoods (or maybe he was just tagging along, on the track of "Love's Body"?), whilst philosopher Albert Hofstadter wandered solitary by day through the same forests, no doubt mulling over his luminous English translations of Heidegger's "Poetically Man Dwells …" And those were just some of our own faculty; the entire 1960s' carnival of wonders came trooping through Santa Cruz, at one time or another. And, in so many words, some of us adopted Bucky. He plainly belonged in the company of those who were going to steer this crazy world back on course. Bucky himself was caught up in this vortex of energy and insight; he knew how to sail with the wind at his back. He came to California often, saying things like, "We can do it. We can make it work. It's now or never, utopia or oblivion." And we young folks listened, enthralled at his three-hour "thinking aloud" monologues, inspired by his mind-bending geometry, astonished at his architectural sleights-of-hand.

As I wrote the foregoing paragraph, it struck me that all of that came to pass barely a third of a century ago. The pattern held, indeed, for quite a while. As far as I was concerned, the party in California went on a good nine years at least, from about 1967 to 1976. But then it seemed to vanish … into tough economic times, New Age hucksterism, political cynicism, yuppies, etc. Or else it splintered into factions: women against men, the Weather Underground versus the New Right, gays against straights, environmentalists versus developers, etc.

And yet in a sense, it is not lost. As Fuller's geometry demonstrates, you miss a lot if you look only at the surfaces. Much of what I have described took place during the dog days of the Vietnam war when Ronald Reagan was playing the role of Governor of California. Soon enough, a grateful nation hired the same bad actor to play President for eight years, all the while supposing they were about to recapture the germinal impulse of the lost American dream. Instead, what they got was a re-run of the values of the 1950s: unregulated 'free' enterprise, fundamentalist Christianity, the patriarchal family, back-to-basics schooling, jingoistic patriotism, etc., all gussied up for the media in pretty 'photo opportunity' packages.

Reagan's Hollywood background completes our scenario (without any need to fast forward to the America of George W. Bush or today's California of 'Gubernator' Arnold Schwarzenegger). It must be admitted that the American dream became, in the twentieth century, mainly a Hollywood production, reflecting not early American history but instead the corrupt and ruthlessly competitive dreams of avarice that went into building Los Angeles. Mike Davis's *City of Quartz* is instructive here; LA is the junkyard of everybody's dreams.[41] Perhaps immigrant Frank Capra's film *It's a Wonderful Life* captures the last moment in the late 1940s when Americans still could see they had a choice between the original dream of a cooperative community and the grubby, grabby, glitzy 'Pottersville' of Jimmy Stewart's nightmare. Today's global village may be no more than Los Angeles writ large, but Los Angeles is just heartless 'Pottersville' sprawling to infinity, and 'Pottersville' was surely Capra's vision of the neighborhood going to Hell. The cherished chestnut *It's a Wonderful Life* still airs every Christmas Eve (as a suicide prevention measure!), but Americans by the 1980s had plainly elected to follow the lead of the Angelenos and spend the rest of their days in 'Pottersville.' Hollywood attempts to revive something of the original small-town American dream in the final decades of the twentieth century succeeded only in taking the notion to quirky *(Northern Exposure)* and politically correct *(Dr. Quinn: Medicine Woman)* points of no return.

Yet even in the Reagan era, some incongruities could never quite be suppressed between the ideological 'image' of flag-waving Americanism, and the 'ghost' of the original vision. Picture if you will that supremely ironic moment of bland Reaganesque incomprehension as he awarded the Medal of Freedom, the highest honor the US government can bestow, to R. Buckminster Fuller – a man who had just had the temerity to suggest to the US Senate that, since national sovereignty was patently obsolete, the United States really ought to set a good example to the rest of the world by desovereigntizing immediately. (A resurgence of the *manitou* of the Iroquois Confederacy? Bucky thought it was his own idea.)

For twenty years, Fuller had been advocating a global, push-button democracy which would give everybody on the planet an up-to-the-minute vote on all the issues of the day. We now take electronic polls at the drop of a hat; why not make these the government, asked Bucky, and reinstitute direct democracy rather than the so-called 'representative' democracy subject to all the abuses we all know all too well? Government by the people, indeed, *by all the people on the planet;* that was Fuller's version, or vision, of the global village. Electronic world democracy may be seen as an extension of Fuller's early perception of 'Continuous Man,' the band of waking consciousness – two-thirds of humanity awake and thinking at any given moment – which circulates, almost unnoticed, around the globe every day. He was always looking for ways to tune in this critical band of wakefulness, which he considered among the most important of the Earth's planetary rhythms. The diurnal oscillations of mindfulness – a feature of the *noosphere,* in Teilhard de Chardin's phrase – would of course have been dramatically evident on an electronic Geoscope, marked by the wide band of sunlight rippling East to West over its seas and continents.

Although almost within reach technically, Fuller's idea of a world democracy never did catch on. Is it any wonder? The staunchest American 'defenders of democracy' don't seem so anxious to play if the game is to include well over a billion highly-motivated Chinese, another billion more potential savvy voters in India, and so on.

And Bucky's more recent plan was even more challenging. In the early 1980s, he had proposed to Canadian Prime Minster Trudeau, and Trudeau in turn to the USSR's Brezhnev, to link up all the world's electrical grids – a move which in Bucky's view would not only make solar power practicable on a global basis (and eliminate the need for oil-fired power stations, or more nuclear and hydro plants), but instantly eliminate both political sovereignties and all the current flim-flam that passes for economics into the bargain.[42] Russian engineers pronounced the plan feasible and desirable; mainstream American science had no comment. After Fuller's death, however, his idea of a planetary electrical grid has been eagerly touted by multinational energy industries as an excuse to build more polluting and destructive power plants in the Third World; a blatant instance of Bucky's ideas corrupted by today's 'globalism-for-profit' crowd. Ignore Bucky's ideas for the free distribution of energy, a cynic might observe, and in twenty years what you get is Enron. But Bucky was no cynic …

All these spinoffs and undercurrents (and even the distortions) indicate something crucial to Fuller's vision: In our synergetic universe, entropy is only apparent. A local system loses energy, granted; things run down and wear out over time. Yet the energy is never irretrievably lost. In the totality of universe, it is eventually harvested and taken up into other systems. Life is anti-entropic. A living pattern – a weather front, a microbe or a mouse, a fern frond, you or me – holds its shape for a time, then disintegrates. We say it 'dies.' But Life goes on.

Did the original American dream really die? And did Fuller's vision, for that matter, die with him?

In many respects, the American dream we hear about these days has become a nightmarish orgy of greed and a moral lesson in the abuse of power. Instead of vision, we have willful mass blindness. Instead of 'livingry,' we have homelessness and ever more weaponry. Certainly Fuller's vision, too, splintered after his death. Various disciples like J. Baldwin, Medard Gabel and Amy Edmonson, or collaborators like Shoji Sadao, Ed Applewhite and Kiyoshi Kuromiya kept parts of it alive. But shortly before Fuller's death, he was introduced to Werner Erhardt, the infamous founder of Erhard Seminars Training, or *est*. So now you can find bits of Bucky's dream marketed like components of any other 'leadership training course,' a New Age version of the now thoroughly debased American dream: success at any cost. Yet Fuller himself claimed he never set out to change people's minds. If he changed the environment by making better artifacts, he supposed, people would just naturally be sensible enough to make up their minds for themselves (and agree with him, presumably). Maybe some of the people, some of the time, but not everybody all at once, and not quickly enough to suit some entrepreneurs of marketable New Age ideas.

So all is not roses, not all of the time, for either the American dream or genuine American dreamers like Bucky Fuller. But all is not thorns, either. Maybe that's what the victory of synergy over entropy really means. A synergy is a whole system, a pattern integrity not comprehensible from a single point of view. The various manifestations of the original American dream we excavated earlier were themselves synergies, integral patterns, knots in complex social, philosophical, environmental and religious nets. Such patterns hold together, sometimes for centuries (when the 'self-reliant' Emerson's house burned down in 1872, his neighbors took up a collection and rebuilt it), then pass away like waves on the sea of time. Fuller describes synergy as one member of a "family of interaccommodative generalized principles" (leverage, for instance, or the inescapable co-existence of tension and compression), a 'family' which functions somewhat like the kinship system of the Iroqouis Confederacy – each principle distinct, none interfering with the operations of any other: "The wellspring of reality is the family of weightless generalized principles." In such a 'family,' each principle implies all the others. The synergy of synergies, the whole concatenation of principles, Fuller called Nature, or in more rapturous moments, Universe: "the aggregate of all humanity's consciously apprehended and communicated … experiences."[43] And Universe abides. It accommodates the rising and falling, the living and dying, the emerging and passing of creatures like us with immense equanimity. "That Universe tolerates our protracted nonsense," Fuller once wryly observed, "suggests significant unrealized potentials."

To return to the Montréal Expo Dome, there is surely no reason why Fuller's own artifacts should be exempt from this dynamic interplay of entropy and syntropy. What appears to be entropy, the dissipation of energy from a local system, must be taken up somewhere, into some other integral pattern. At least Bucky himself would see it so. And now that time, fire, ice and wind have ripped through its very fabric, where did the energy – or, indeed, the integrity – of that vast structure go? Parts have been rebuilt, and the shell repainted. Bits of the whole structure fell back as ash into the native American soil and landforms, but other parts – above all, the intact skeleton of the whole – can still touch the eyes and minds and hearts of all who lay eyes upon it. In a sense, that energy has gone into you and me, as well as into Robert Duchesnay's starkly enigmatic photographs. Both Harry Kroto and Richard Smalley, joint discoverers of Buckminsterfullerene, had visited the Montréal Dome during its Expo heyday, and Kroto had built a geodesic playdome for his kids. The two say they hit upon the soccer-ball structure of 'Buckyballs' by recalling Fuller's domes, which show plainly that in order to bend the surface of a flat hexagonal matrix all the way round into a sphere, you have to take out ten (36°) pentagonal tucks. Bucky had solved their problem, years before it arose, and they honored him by naming the new molecule after him. But this is only a single instance. Surely Bucky's shapes have inspired plenty of other people as well, in many and even more subtle ways.

* * *

For decades, Bucky's domes seemed to be 'the shape of things to come.' They were even used in the color sequel to the H. G. Wells' film of that name, as well in other sci-fi films like *Silent Running* to symbolize the farthest reaches of human dwelling. Once upon a time, the first Sumerian and Egyptian attempts at civilization tried pyramidal and hierarchical forms – not only for their monumental architecture, but for the organization of their totalitarian societies. In our own day, we've tried square and cubical and gridiron forms, and a more egalitarian society mirrored these shapes. For nearly a century, the machine age gave us merely the vertical projection of these forms, pulled up into the skyscraper and the new corporate and bureaucratic hierarchies these shapes tend to accommodate. Bucky wanted to take the industrial tooling of architecture a step further, to give us a spherical or at least hemispherical model for building, and perhaps also a more well-rounded notion of human dwelling within the living biosphere.

In a sense, all knowledge is local, and partial – bound by language, culture, tradition and landforms. Bucky might be reluctant to admit it, but his own grand, globe-girdling schemes had detectable New England roots. Yet every local truth is equally bound to the whole Earth, and every partial discovery seeks the widest possible stage upon which to play its part. Fuller never consciously constructed his domes to fit into their particular settings the way Frank Lloyd Wright designed his houses. (In the year 2000, Wright's beautiful 'Falling Water' home was found to be leaking and falling down; more attention to *structure* might have headed off the seven-million-dollar restoration effort.) Geodesic domes, by contrast, were usually designed in the abstract, with more respect for sheer principle than for setting.

It is therefore more than a little uncanny to travel through Asheville, North Carolina, where Bucky built the first geodesic domes at Black Mountain College in the late 1940s, and discover a place where domes actually look like the landforms – or rather vice-versa. Asheville straddles the juncture of the Blue Mountains and the Great Smokeys. This is Cherokee country, the old arch-rivals of the Iroquois Confederacy. And here, wherever you turn, you find little dome-like hills above you and below you, the result of the two rippling mountain chains merging in steep little troughs and weathered crests. Here, oddly enough at their very birthplace, Bucky's round houses would seem to leap right out of the surrounding landscape as the most obvious and adaptive human variation on a natural theme.

The strong spirit of place notwithstanding, there are no domes remaining here, no artifacts at all from those first experiments in geodesics. When I visited, the site of Black Mountain College had become a conventional boy's school. Yet it was for a few memorable years one of the most remarkable American efforts to create a free-form open university. During the late 1940s and early 1950s it was presided over by the titanic Charles Olson, with occasional faculty like Fuller, Marshall McLuhan, Norman O. Brown, 'musician' John Cage, filmmaker Arthur Penn, or choreographer Merce Cunningham, and attended mostly by unconventional, creative types like poets Robert Creely, Denise Levertov and Robert Duncan. Years on, one could still feel the reverberations of that experiment in the hopes and dreams of faculty members who eventually tried to recreate the Black Mountain experience at places like UC Santa Cruz, or more recently Naropa in Colorado. Something similar may be said of Fuller's domes. Not a trace left at the site of their origin, not even a commemorative plaque (in a town which has fastidiously preserved novelist Thomas Wolfe's house as a civic shrine), and yet powerful seeds were flung from this place, and proved so hardy that by now over 350,000 geodesic domes (if you count playdomes) have sprouted up, a sprinkling on every continent – even Antarctica.

Physical structures deteriorate over time, it is true. Entropy rules the physical. Yet synergy, as Fuller saw it, is patterned integrity – timeless, weightless, indestructible: "We are now synergetically forced to conclude that all phenomena are metaphysical; wherefore, as many have long suspected – like it or not – life is but a dream."

In so many words, integrity, and maybe only integrity, endures the bite of time. Once again, it is a matter of principle.

I suspect Bucky Fuller himself would not be overly exercised about his own 'legacy.' He would, however, be intensely interested to see whether the human race manages to survive the twenty-first century:

Though humans are born equipped
To participate
In the supreme function of Universe,
This does not guarantee
That they will do so.
. . .
They must first discover the overwhelmingly superior efficacy
Of mind, as compared to muscle.
And humans also must discover
That the physical,
Which they tend to prize as seemingly vital,
Is utterly subordinate
To the omni-integrity of metaphysical laws
Which are discoverable only by mind.
. . .
Only if man learns in time
To accredit the weightless thinking
Over the physical values,
In a realistic, economic and philosophical accounting
Of all his affairs,
Will the particular team of humans
Now aboard planet Earth
Survive to perform their function.[44]

II. Refractions

Dwelling Machine

Toward the close of Robert Snyder's film, *The World of Buckminster Fuller* (1971), Bucky puts his vision for technologically-enhanced human dwelling into a single telling image. He has been enthusiastically describing NASA's 'black box,' a self-contained human sustenance kit (including food, water, electricity, plumbing, etc.; all the comforts of home) for life in outer space. Although it might initially cost $20 billion in prototype, Fuller foresees it as eventually a $1,000/500 lb. package no bigger than "a good-sized suitcase" for setting up house here on Earth. Add to this a "geodesic space umbrella," and humankind might well enter a new age of living lightly upon the Earth:

> *You have in effect the man with his umbrella and his briefcase able to go anywhere in Universe – pick a beautiful spot! – and live at a very high standard of living.*

I'd like to linger with this image because I suspect it lies behind many of Fuller's artifacts and inventions, and may provide us with a kind of golden thread running from the beginning of his career as a builder to his later years as world-around 'comprehensive anticipatory designer.' What does the image include? A lone man (is that sexist?), with briefcase and umbrella (Bucky on his way to a lecture?), able to pitch his geodesic tent wherever and whenever he likes. Forget the inessentials – that the briefcase weighs a quarter ton, that the umbrella is a geodesic canopy 30-40 feet in diameter. Forget the fact that NASA's spin-off contributions to 'livingry' here on Earth so far have tended to range from the trivial to the ridiculous (dustbusters, ingestible toothpaste, battery-operated gloves and boots, invisible braces, liquid-cooled garments, rechargeable electric footwear, self-adjusting sunglasses, sports bras and therapeutic scalp coolers, for instance). Stick with the image, a futuristic idea attired in curiously antique apparel; all that's missing is a bowler hat equipped with inverted satellite dish. In some respects it is an undeniably exhilarating image – mainly for the absolute autonomy it promises, a freedom not only to move but to live wherever and however you like. Given that the 'black box' would presumably include a version of the self-contained 'Dymaxion Bathroom' Fuller invented in the 1930s, one is even encouraged to contemplate liberating one's bowels from the coils of a central plumbing and water system. 'Dymaxion,' by the way, was Fuller's trademark back in the 1930s and 1940s, coined by a publicist from bits of words he picked out of Fuller's idiosyncratic vocabulary, presumably combining 'maximum' (efficiency?) and 'dynamic' (systems?) with a strand of 'tension.'

But back to our image, we have also to ask: what is missing from this picture? Most obviously, there is no community here, not even any family. And surely more questions are raised than answered about land, property, ownership, etc. Above all, as native peoples would be quick to point out, there are no *roots* to this kind of dwelling, no permanence envisioned and no stewardship of the land implied. This deracination is a refractory issue we shall have to attend to a bit further on.

For now, it may suffice simply to outline some of the steps by which Bucky worked his way through customary building practices, conceptual blocks, and recalcitrant materials over the course of half a century to realize this ideal of the man with dwelling-kit briefcase and geodesic umbrella, free to set forth and settle down wherever in the wide world he chooses to be.

It began after he had spent much of the 1920s refining his father-in-law's patents on the Stockade bricks structural system, which incidentally became the basis for building with acoustical wall and ceiling materials. In 1927, Charles Lindbergh had only just flown the *Spirit of St. Louis* nonstop from New York to Paris in 33.5 hours, but Fuller was already thinking about great-circle polar routes (five years before anybody else mapped them) and designing multiple-deck tower apartments, with expanded wire-wheel construction. These were to be flown to the North Pole by Zeppelin and erected in a single day. He called them "stepping stone, world airline maintenance crew environment controls," anticipating a "one-town world."

Although never built, these so-called "4D multi-storey dwelling units" were the prototypes for the Dymaxion Dwelling Machine, the only complete houses Fuller did actually build. At the time, according to Bucky and Robert Marx in *Dymaxion World,* he listed the advantages of his "multi-storey units" as:

> *Completely independent power, light, heat, sewage disposal; 12 decks average 675 sq. ft. each; all high in air – above dust area, etc.; all furniture built in; swimming pool, gymnasium, infirmary, etc.; as free of land as a boat; time to erect – 1 day; fireproof.*[1]

These he compared with the limitations of the conventional house –

> *[a] tailor-made archaic contraption with little or no sunlight; jiggle and she'll bust; tied down to city sewerage system, the coal or oil company – the utility; six rooms average 225 sq. ft. each; down on ground, subject to dust, flood, vermin, marauders; no pool, etc.; no structural improvement in 5,000 years – if anything, retrogression; time to erect – 6 months; not fireproof.*[2]

It took another 18 years and the resources of the Beech Aircraft Corporation – which Fuller wanted to see retooled from wartime production to 'livingry' – to build a pair of prototypes for the scaled down single-family aluminum-shelled Dymaxion House in 1946, but the industry Fuller envisaged never materialized. Big-hearted as ever, Bucky offered his patents on the 4D house *gratis* to the American Institute of Architects. They huffed and they puffed and they passed a resolution putting the idea down: "The American Institute of Artchitects is opposed to any kind of house designs that are manufactured like-as-peas-in-a-pod." Never mind, as Hugh Kenner notes in relating the anecdote, that peas in a pod are a marvel of design engineering. Mass-produced houses were supposedly unthinkable, and maybe Fuller's generosity in just giving his designs away even more so.

But Fuller had been thinking seriously about applying industrial techniques to housing for a long while. It seemed to him preposterous that weaponry and ship-

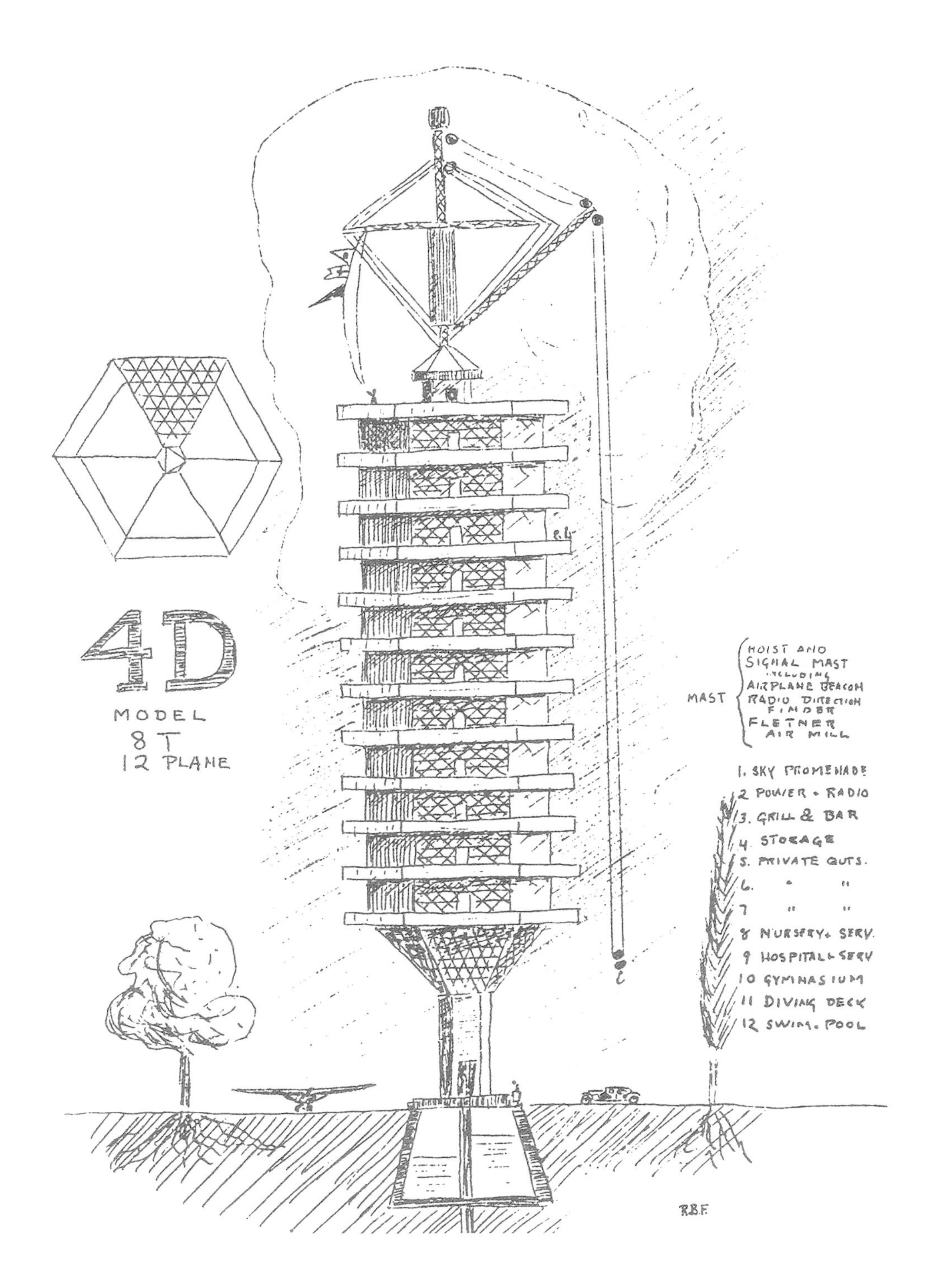

Multi-deck tower apartments; wire-wheel construction (1927).
(Estate of R. Buckminster Fuller.)

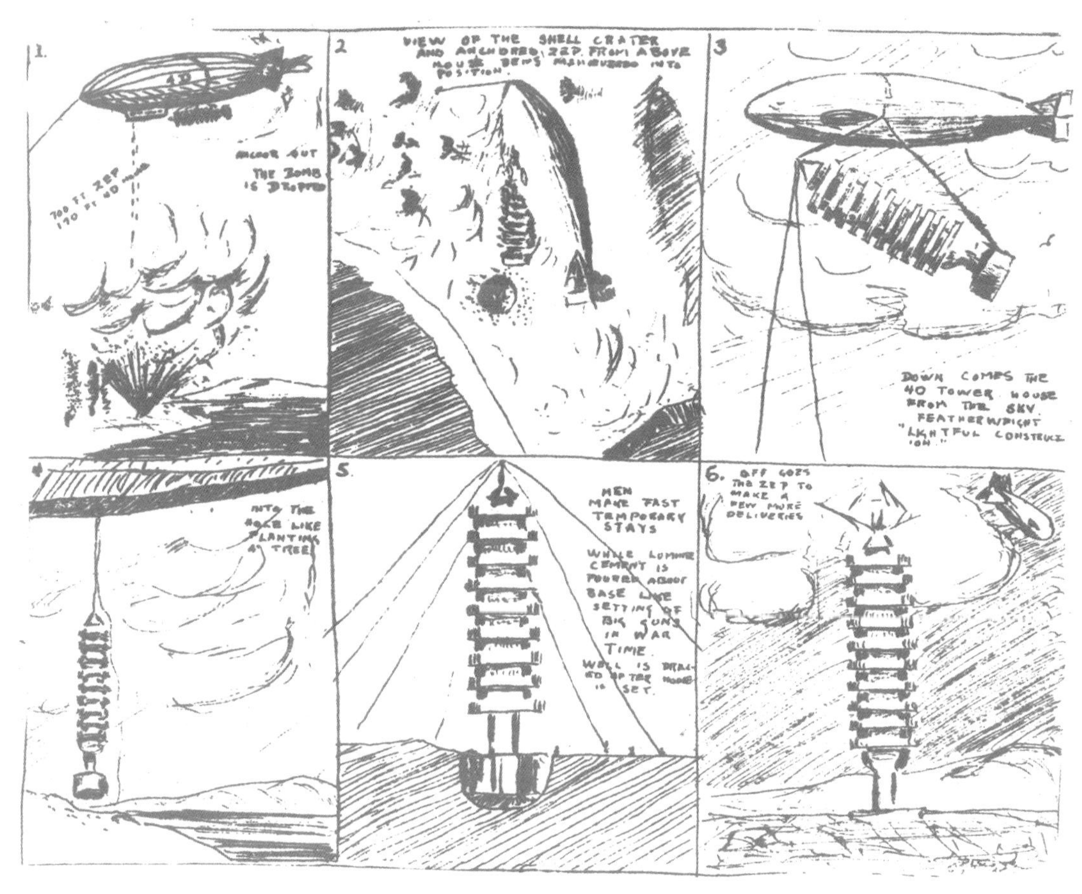

4D dwellings – Delivery by Zeppelin and North Pole construction (1927). (Estate of R. Buckminster Fuller.)

building and the aircraft industry should progress so rapidly whilst the very structures which nurtured and sustained human life stagnated somewhere back in the late stone age. The impulse might well have gone back to his time in the Chicago tenements, where his first daughter Alexandra died of a combination of preventable diseases, and his wife Anne spent most of the day "chasing yesterday's dirt." Bucky reasonably surmised that a stunted environment produced stunted human personalities. Give the child a good environment, he countered, and it will develop to its full capacities. Simplistic, perhaps, but compelling. The aim is to free people from all that outmoded housing required of them – a twenty- or thirty-year mortgage, immobilization in centralized plumbing and electrical systems, a minimum of space for a maximum of cost, etc., etc. This, quite clearly, the superficial functionalism of the Bauhaus, for instance, did not do or even aim to do. The so-called International Style settled for only the facade of utility: "They never looked at the plumbing," Bucky declared. And so he went his own way. In 2001, the Ford Foundation managed to collect and painstakingly reassemble bits and pieces from both prototypes in order to restore a single Dwelling Machine for the Henry Ford Museum in Dearborn, Michigan.

Was the sticking point really the notion that housing could be mass-produced? Why should it be? Tenements were already mass-produced, as were many of the Sears & Roebuck kits for the sturdy mail-order Victorian houses that still grace the streets of

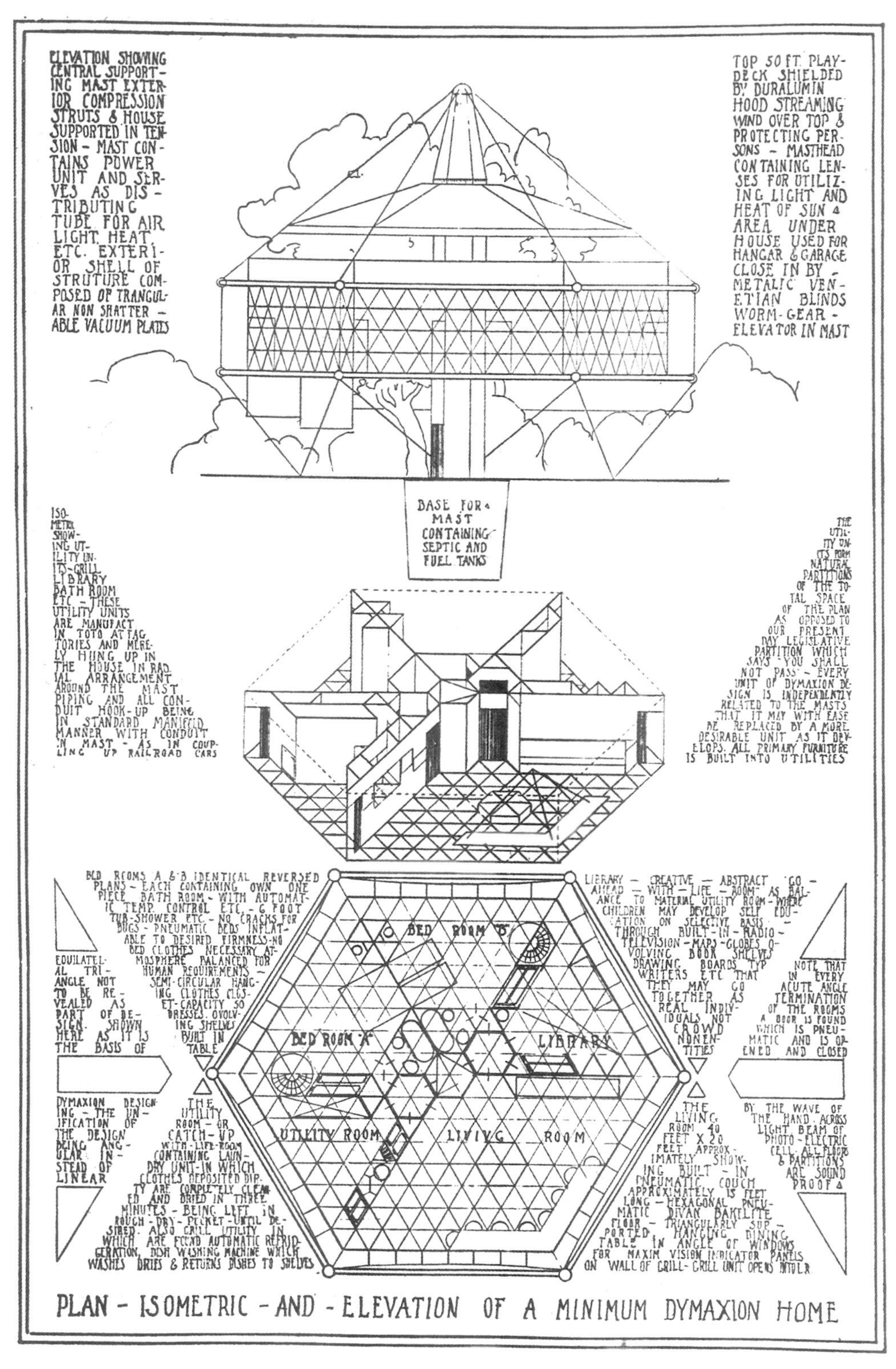

Plan of a minimum dymaxion home. (Estate of R. Buckminster Fuller.)

Western cities like San Francisco (the Haight-Ashbury is full of them). Lumber and piping were mass-produced; nails were mass-produced; shingles and fittings and entire floor plans were already mass-produced. No, kits and mass-production were acceptable – for low income housing, at the very least – if not the preferred work of architects looking for hefty commissions.

Had they been produced, I suspect that Fuller's Dwelling Machines would have met unexpected, even illogical, resistance to the logical next steps: machine-tooled housing, and metal structures. Never mind if it's cheaper, more efficient, lighter in weight, more durable and all those good things Fuller extolled. Who wants to live in a glorified tin can? Fuller's 1940 'Dymaxion Deployment Units' (a quick half-step toward the Dwelling Machine, 7,500 of which were built for the US Army on the design of metal grain silos) even *looked* like tin cans with portholes, although fairly comfortable inside. As a matter of fact, Fuller's use of metals was essential to his vision. It had two roots – one he used often to illustrate the very concept of synergy, the other a necessity of structures which rely mainly on tension rather than compression.

In the first place, Fuller considered metallurgy and the advantages of alloys to be virtually a paradigm for human evolution and progressive 'mastery' of the elements. Take an instance Bucky cited often, the alloy chrome-nickel-steel. The combined tensile strength of the three metals taken separately is 260,000 pounds per square inch. But as an alloy, their tensile strength increases to about 350,000 p.s.i. That unexpected increment, the extra 90,000 p.s.i., is sheer synergy – the behavior of the whole utterly unpredicted by the behavior of its components taken separately. The gain in efficiency is obvious.

Metallurgy also exemplified what Fuller liked to call 'ephemeralization,' the ever-progressing know-how which enables thinking humans (using Mind, which discerns overall patterns and principles, and not just Brain, which merely stores special-case instances) to do "more and more with less and less." This trend was his answer to the pessimism of Malthus, and a prime gradient upon which he predicated his own optimism about progress and the betterment of human life.

The second reason to turn to metals was precisely their high tensile strength. Fuller's constructions deliberately reverse many of the structural biases inherent in the way Western peoples have gone about building things throughout history. Whether gathering rocks from a field and piling them up into a house, or clearing the field of timber later used to build the house, we Westerners have always relied upon the weight of compression (push) to build things – lintels nailed atop posts; bricks piled atop bricks; little boxes jammed together within bigger boxes, etc. Fuller took his lessons from Nature, by discerning that Nature builds her large structures (from the spider's web to the solar system) by relying much more upon tension (pull); tensile strands joined at knots of compression.

The Dymaxion 'House on a Pole' was really an expanded pair of interlocked wire-wheels; the hubs were elongated into a mast which contained all the plumbing, electrical and maintenance functions, while the floors and ceilings of the house were suspended by cables from the upper boom. Water, septic and fuel tanks he tucked neatly away in the foundation, the supporting structures were pneumatic (thin aluminum tubes filled with air), and the walls transparent plastic plates. The house was to be assembled top down, the construction sequence illustrated in an elegant model worked up by Fuller

The Wichita House, 1945. (Estate of R. Buckminster Fuller.)

The remains of the Wichita House. (© Robert Duchesnay, 1990.)

Restoring the Dymaxion Dwelling Machine. (© Robert Duchesnay, 1992.)

and some of his ever-enthusiastic architecture students. Interestingly, as Kenner notes, when Fuller first designed the Dymaxion House in 1927, very few of its appurtenances (washer-drier, temperature controlled ventilation, dishwasher, radio-television receiver) even existed. Most Americans were still living in houses "pasted, piled and tacked together ... without even bathrooms, toilets or sewage disposal."[3] He was anticipating not just a nicer house, but an industry with an expert research-and-development team devoted to refining such things. You do not build thousands of miles of roads, manufacture tons of metal alloys and pump lakes of liquid fuel to produce only a single automobile. Nor, Fuller reasoned, should you build houses that way. His premise was simple: the time has come to extend industrial strategies to housing. In the end, what Fuller was trying to engineer – all by himself – was an entire industry which would provide you a 'dwelling unit' the same way (and at nearly the same cost!) as the phone company provides you a phone:

> *In a few years you may be able to walk into a telephone company booth one morning and ask them to put up a dwelling device on a certain mountain that afternoon. When you're through with it, you will call the company and ask them to remove it. The company will have your environment control waiting for you wherever you want it at low cost for dream high standards.*[4]

Here yet again, we glimpse our man with briefcase and umbrella – this time provided for by the housing industry equivalent of fast food outlets. To get there, he designed and built for the 1934 Chicago World's Fair his three-wheeler Dymaxion Omni-Medium Transport, which was not – he insisted – an automobile, but only the aerodynamically designed ground-taxiing fuselage for a vehicle able to take you over land or water or through the air to and from your Dwelling Machine (which had a parking place already designed for it). Yet even with two prototype 4D houses actually built in 1946, the industry failed to take off, at least partly because Bucky seemed to be temperamentally unsuited to the games venture capitalists play.

The way his old friend Alden Hatch tells the story, he simply had no stomach for what he considered the inherently corrupt and corrupting practices of big business. He had backers, he had 3,500 orders, he had the means to raise the estimated $10 million needed to get production rolling, etc., but at the last minute he kept revising the plans, stalling on any proposed shortcuts, insisting on a full-fledged delivery system, and so forth, until Beech Aircraft pulled out and frustrated associates finally left Fuller and his self-contained house to their own devices. The Dwelling Machine wasn't ready, he kept repeating. More likely, profit-driven American business wasn't ready to undertake the entire altruistic service industry Bucky had in mind. This was not the only time he walked away from a possible fortune in order to keep the integrity of his vision intact. The War had demonstrated what America's tremendous industrial capacity could produce when pressed into service for weaponry. Bucky wanted to rededicate industry to 'livingry,' to harness those vast reserves of power and ingenuity to the service of life. In later years, he used to claim that "making money and making sense are mutually exclusive." At any rate, from this point on he concentrated exclusively on the envelope, anticipating that industry – and later NASA – would soon enough perfect the 'black box' or 'autonomous dwelling device.'

Fuller went on from Wichita to Black Mountain College and there started building

Fuller and Dwelling Machine model. (Estate of R. Buckminster Fuller.)

domes, as we have seen. Geodesic (great-circle) structures – actually first invented in 1922 by Dr. Walter Bauersfeld as a lightweight icosahedral steel lattice for the rooftop dome of his first Zeiss planetarium in Jena, Germany[5] – emerged for Bucky directly from his own work on the Dymaxion Air-Ocean world map. His first domes mapped the great circles of the vector equilibrium and icosa patterns he was using to subdivide the globe; later he set about subdividing the twenty icosahedron triangles themselves, while allowing the great circles patterns to remain almost invisible in the tensional network holding the whole together. The domes were his one great commercial success. As I write today, the Ethiopian Parliament is meeting under one of them in Addis Ababa. If it were not for the domes, he admitted from time to time, you probably would never have heard of Buckminster Fuller. And if it were not for the income from his patents on the domes, he would not have become the global lecturer of his later years, tirelessly promoting his own technologically-enhanced vision of how to 'make the world work.'

The domes were in a sense only the symbols for Fuller's larger ideas of human dwelling on Planet Earth. Just as Fuller first derived his geodesic math from work he was doing on the Dymaxion Air-Ocean World cartographic projection, so the Expo Dome, as we have seen, was intended to enclose a Geoscope, the computerized space-age version of the original air-age Dymaxion Map. Although not interchangeable, Whole and Part can include one another, and even turn into one another. According to Fuller, you begin with Whole Systems – the Universe (Nature), and specifically planet Earth – and from there, you learn to focus in on local instances. Reciprocally, this 'local problem-solving' – which he claimed to be Man's prime evolutionary function – led you (not directly, perhaps, but by great circle routes) back into the Big Picture again, looking for the principles to apply to that particular situation. He called this 'valving' the Universe into special-case applications. In the same way, his domes were mainly (though not always) derived from icosahedral hex-pent coordinates. He called the icosa "the prime dwelling valve of Universe," a form pulled out of the isotropic vector matrix which gives maximum space enclosure for minimum investments of materials and energy. This reciprocating movement from deduction to induction and back again to first principles tended to run against the grain of mainstream science, still wary of the generalizations of the medieval schoolmen and dogmatically tied to solely inductive, analytical methods. By starting with the Whole, and returning to it ever thereafter, Fuller arrived at places and structures you simply could never get to by adding up the bits and pieces in hopes of one fine day achieving the Big Picture.

By the mid-1960s, Fuller was proposing to dome over entire cities, creating huge "environment control devices" which he claimed would offer "the enormous advantages of the extraversion of privacy and the introversion of the community." Now whatever he may have meant by this, we've run into that word "community" missing from the image of the man with the briefcase and umbrella. But how does Fuller envisage communities? In the 1970s, when his design for a floating community called Triton City emerged, or his big domed stadium proposal (called "Old Man River") for low

Proposed dome over Manhattan. (Estate of R. Buckminster Fuller.)

income housing in St. Louis, it appeared that he had in mind communities as some sort of gigantic composite version of his 'individual' dweller – insulated from the vicissitudes of weather, isolated from one another geophysically, entirely autonomous, and sometimes even mobile. He described, for instance, "a great tetrahedronal city, to house a million people," and sketched it provocatively floating in the middle of San Francisco Bay. And as a final challenge, he pointed out that when geodesic domes – the only manmade structures that get stronger as they get larger – grew to a half-mile or more in diameter, they would weigh so much less than the air inside them that they'd simply float skyward unless tethered. The weight of human "communities" incorporated into these spheroid sky cities, he maintained, would be "negligible." Thus the Montréal Expo Dome was no more than a baby 600-ton bubble, a prototype for these vast "environment control devices." He even designed an "Octa Spinner" to mass-produce the weave from space stations (which are themselves constructed today with 'his' octet truss as their basic structural component). Whether feasible or not, desirable or not, these structures all have one important thing in common. They seem to encourage humans to live more lightly on the Earth, even to clear right off the land masses for a time, as if to permit the stressed eco-systems of the planet to regenerate themselves. This is the positive reading: "… that man may be able to converge and deploy around earth without its depletion."

On the negative side, Fuller's notion of the communities meant to fit into these mega-structures seems rather thin and unconvincing. He envisioned communities almost as composite machines, with a multitude of elements, working parts and interactions, all the components of which he would happily design for you. But totally planned communities are not most people's idea of utopia, despite the prevalence of retirement 'communities' for dis-used old people and the recent efforts of the Rouse Corporation (which gave us the EMAC, "Enclosed Mall, Air-Conditioned") and others to design, build and run planned communities with walls and gates and security guards around them to protect the wealthy from the deteriorating urban milieux their own profiteering has created. These self-imposed prisons resemble the picturesque but stifling 'Village' in Patrick McGoohan's celebrated *Prisoner* TV series of the late 1960s: sterile, petty, conventional totalitarian states full of bland faces, all smiles and no soul.

However all this may be, Bucky certainly did not foresee what would happen with dome architecture, except for the very large constructions (Ford Rotunda, American Society for Metals, Union Tank Car, Kaiser Domes, etc.) and pioneering prototypes (Radomes, Skybreaks, Seedpod Foldables, Fly's Eye, and paperboard domes, etc.) which he and his architectural partner Shoji Sadao actually supervised themselves. Fuller expected his geodesic domes to become overnight a planet-wide housing industry which would mass-produce cheap, efficient 'Dwelling Machines' by the millions. He expected the 'black box' autonomous dwelling kit as imminent terrestrial fallout from the space program.

What happened instead quite probably surprised him: the re-runs of *Star Trek* lasted longer than the nation's wholehearted commitment to the space program, and it turned out that the people who actually built and lived in domes did so to express their choice of an alternative lifestyle more or less detached from the mainstream commodity culture. They appealed to the low-budget do-it-yourselfer (who tended to build in wood), not to the industrial manufacturer in search of profits. They appealed to hippy communes

Boxed in

Not a single geodesic dome home was destroyed in the California earthquakes of the 1990s, nor in all the hurricanes that have devastated Florida, Louisiana, and the mid-Atlantic states since global warming increased both their frequency and intensity. Conventional box buildings were easily leveled by a little shake or a big breeze. Yet when people rebuild, up will go the boxes again.

Why?

We have conditioned ourselves to see *life in a box* as normal and natural. It is the very shape of Western culture. We are born into a box (cradle, crib and playpen), we spend every day and every night in a box (home, office, automobile), and we generally insist that our remains be buried in a box.

Wherever we westerners leave our mark, it seems to be at right angles to the rest of the Universe. Before the European hordes overran the North American continent, to stick with our example, the inhabitants managed to live happily in triangulated structures (tipis) or round houses (hogans, igloos, etc.). Now, everywhere you turn, boxes piled atop boxes, each of them numbered, mortgaged, insured, and taxed to the hilt. Until the next earthquake, anyway …

So at home are we in our box that we scarcely even notice it. Western sports and politics, our legal systems, economics and government, all share the same 'framework.' The square is everywhere. We squeeze everything into the neat compartments – us/them, right/left, black/white, yes/no – with which we would divide and conquer reality.

We have squared off against the Earth herself, as if to repudiate her roundness. And all those rifts dividing the human family? Merely lines we ourselves have drawn.

There are alternatives. Bucky Fuller spent his life demonstrating them. His 'radomes' in the Arctic weathered temperatures of 50 below zero and winds up to 150 mph, yet were still in good enough shape when decommissioned half a century later to be given to local Canadian communities in need of durable community shelters.

But until such alternatives are taken seriously, we must conclude that here is a civilization which has quite literally boxed itself in.

Baton Rouge Dome. (© Robert Duchesnay, 1990.)

more readily than to single families, maybe not only for their 'cosmic' embodiment of Nature's principles – "Every vertex a mandala," as Hugh Kenner once observed of the garden restaurant dome in Isla Vista, CA – but due to pragmatic constraints like leaky roofs, adverse zoning codes, or an open structure and sound-in-the-round acoustics which tend to minimize the privacy family members so often crave from one another. Maybe because he was himself such a maverick, Fuller's ideas have often tended to appeal to non-conformists who in turn applied them in their own distinctive ways.

Many commentators seem to think Bucky's ideas stand or fall by the success or failure of the dome-building industry. I beg to differ. I find his fundamental re-envisioning of technology as 'what Nature does,' and his consequent deployment of 'Nature's coordinate system' (not solely in geodesics but in many other ways as well), to be initiatives more far-reaching – and even more durable – than the domes themselves.

All of which ironically brings us back – by one of those invisible great circle routes, no doubt – to the moral and spiritual climate of New England.

Ambivalent Centennial

During August of 1858, Ralph Waldo Emerson set out in the company of ten other townsmen – doctors, scholars, clerks, he notes – on foot and by canoe to experience the still-wild Adirondack Mountains in upstate New York, deep into what was once Mohawk territory. He kept a journal in blank verse, the early stanzas recording his joy at rediscovering Nature untrammeled.[6] "So fast will Nature acclimate her sons" that he and his fellow pilgrims ("Chaucer had no such worthy crew") were soon rejoicing to leave behind the ways and wiles of cities and towns:

Lords of this realm,
Bounded by dawn and sunset, and the day
Rounded by hours where each outdid the last
In miracles of pomp, we must be proud,
As if associates of the sylvan gods.
We seemed the dwellers of the zodiac,
So pure the Alpine element we breathed,
So light, so lofty pictures came and went.
We trode on air, contemned the distant town,
Its timorous ways, [and] big trifles …

Their senses enlivened, the townsmen learned to read again the signs and portents of sky and cloud, deer and eagle, midge and mosquito: "O world!," exclaimed Emerson, "What pictures and what harmonies are thine!" One observes that in the very act of throwing off the conventions of 'civilized' life, Emerson also managed for once to liberate himself from the artificial metrical conventions which so often contorted his efforts at poetry (exceptions like "Brahma" only proving the rule). Had he kept this up, Emerson's own poetic 'essays' might more often have matched the supple grace and freedom of his prose.

Judge with what sweet surprises Nature spoke
To each apart, lifting her lovely shows

To spiritual lessons pointed home.
And as through dreams in watches of the night,
So through all creatures in their form and ways
Some mystic hint accosts the vigilant
Not clearly voiced, but waking with a new sense
Inviting to new knowledge, one with old.

But just as the travellers began to sink snugly into the bosom of Mother Nature – Emerson comparing them to explorers or hunter-gatherers of old – something momentous happens. A newspaper, "big with great news" of the first transatlantic telegraph cable, arrives from the wider world. The atmosphere of Emerson's journal abruptly changes; now he exults in human ingenuity, while pristine Nature fades discreetly into the background:

With a vermillion pencil mark the day
When of our little fleet three cruising skiffs
Entering Big Tupper, bound for the foaming Falls
Of loud Bog River, suddenly confront
Two of our mates returning with swift oars.
One held a printed journal waving high
Caught from a late-arriving traveller,
Big with great news, and shouted the report
For which the world had waited, now firm fact,
Of the wire-cable laid beneath the sea,
And landed on our coast, and pulsating
With ductile fire. Loud, exulting cries
From boat to boat, and to the echoes round,
Greet the glad miracle. Thought's new-found path
Shall supplement henceforth all trodden ways,
Match God's equator with a zone of art,
And lift man's public action to a height
Worthy the enormous cloud of witnesses,
When linkéd hemispheres attest his deed.

At this point, all thoughts turn back to the doings of men, exemplified by the Atlantic Cable – which failed by the way, in October (for lack of proper insulation), and only got properly underway in 1866. Yet this might well be considered the first major link in today's information 'superhighway' – "Thought's new-found path" – and Emerson seems to see it as a kind of cultural correlate, or precipitate, of all the meaningful interconnections he and his fellow "dwellers of the zodiac" had been perceiving in Nature. Curiously, the intelligible shock wave from the telegrapher's submarine spark comes to the company of travellers first by "shouted" word of mouth, and then by "printed journal," inverting the entire history of media from orality to literacy to electricity. Emerson's paean now absorbs the news by turning in amazement to "This feat of wit, this triumph of mankind … Urging astonished Chaos with a thrill to be brain, or serve the brain of man" –

Mind wakes a new-born giant from her sleep.
Twirl the old wheels! Time takes fresh start again,
On for a thousand years of genius more.

At such a triumphal moment, the intrepid travellers no longer wish to linger in the wild. A final image sets out to reconcile, if not indeed to subordinate, the bounteous beauties and wisdom of Nature to the electric spark of human genius:

… So in the gladness of the new event
We struck our camp, and left the happy hills.
The fortunate star that rose on us sank not;
The prodigal sunshine rested on the land,
The rivers gambolled onward to the sea,
And Nature, the inscrutable and mute,
Permitted on her infinite repose
Almost a smile to steal to cheer her sons,
As if one riddle of the Sphinx were guessed.

Emerson was able, as we today are not so easily able, to "Match God's equator with a zone of art," to hold in dynamic balance "inscrutable … mute" Nature and "the glad miracle" of Culture, rustic wisdom and technological innovation. It's as if he is simply turning his head – nodding now toward "the sylvan gods" of Nature and the primeval past, now toward the "ductile fire" of technology and the "new-born" future. Others, even in his 'Golden Day' of American letters,[7] seemed already to be shaving off in the one direction or the other. Whitman did manage to embrace both Nature and Culture, but only as reflections of his much vaunted 'Self'; Thoreau and Melville leaned toward Nature – 'pure and simple' in the case of Thoreau; raw, elemental and even inhuman in Melville; – whilst Hawthorne could only draw his Gothic portraits from tortured townlife. But Emerson neither saw nor felt any contradiction in casting off the trappings of civilization and singing the praises of Nature, only to find himself celebrating new facets of human genius from the depths of the Adirondack wilderness: "We flee away from cities, but we bring the best of cities with us."

* * *

Since the time of Emerson's Adirondack interlude, the rift has deepened between the twin facets of his vision. As a culture, we are now profoundly ambivalent about our own technological artifacts. Since industrial manufacturing techniques were first applied to mechanized warfare in World War I, and grotesquely exaggerated by the Bomb and crematoria of World War II, we have seen all too much of the dark side, the eclipse of Nature and the distortion of human nature, latent in our technologies. 1995 marked the Centennial of the two great 'holistic' theorists of technology North America produced during the twentieth century: Bucky Fuller, of course, but also Lewis Mumford, who might well be considered Fuller's arch-critic … as much for the ease with which he was able to dismiss Fuller as for what he actually deigned to say about him. The coincidence signaled a deeply ambivalent Centennial. The divergences between these two thinkers highlight an intractable split in contemporary attitudes toward technology. In their work, the rift barely discernible in Emerson becomes a chasm.

Lewis Mumford lived most of his adult life on farmland in the rolling hills of Amenia, New York, in Dutchess County northeast of New York City (famous, or infamous to some, for Woodstock and Millbrook). While southeastern New York State may only border on today's New England geographically, Mumford was at least as much an inheritor of the quintessentially New England spirit as Bucky Fuller. The passing years have only magnified Mumford's stature as a pre-eminent American architecture critic for half a century, as well as the importance of his ground-breaking work as an historian of technics. After a stint in the navy during World War I (much like Bucky's), Mumford began his career promoting the 'organicist' ideas of Patrick Geddes – the Edinburgh biologist and city-planner who collaborated with Ebenezer Howard in originating the 'Garden Cities' movement in England – but beyond their common emphasis on regionalism, he soon shook off any remnants of discipleship. His first book, *The Story of Utopias* (1922),[8] outlined with prescient accuracy the fatal flaw in all totally 'planned' communities. On paper, utopias appear to be reasonable alternatives to the status quo. In practice, however, they tend to have no room for alternative utopian visionaries, i.e., for anybody whose own views might be at variance with the original plan. Which is why utopias – from Plato's *Republic* to the many utopian experiments on American soil – all too easily and too often tend to become totalitarian.

Much the same impulse lay behind his first book on American architecture, *Sticks & Stones* (1924),[9] which formed the basis for his later and more famous *The City in History* (1961).[10] Here Mumford extolled 17th century New England towns for continuing the medieval polytechnic tradition, where every craftsman was a master and an autonomous source of innovation, working in the free air of interdependence we earlier underscored in our archaeology of the original American dream. He did not go so far back into the native soil as to re-discover the *manitou* of the Iroquois Confederacy, but he did share much the same admiration for the 'socialist' virtues of early American communities expressed by Emerson and de Tocqueville. He, too, saw the cooperative communitarian dimensions of town life as the root of the American spirit in early New England, and decried its fatal erosion by both the mercantilism of the great commercial cities, and the slash-and-burn profiteering of the pioneer mentality. In his later books on technics and civilization, he took a hard look at the irrational streak (wars, persecutions, intolerance) accompanying the most rational plans for civilization from its Mesopotamian origins to the present. In the American context, he first tried to inflect, then began to resist the increasing domination of architecture (since the nineteenth century) by architects and industrial engineers, and of community life by social 'engineers.' The only sorts of town plans he endorsed maximized green space and open areas for interaction, indeed, a variety of 'unplanned' spaces in which people might freely associate. Otherwise, the organic growth of community could only be cramped by the limitations of the artificial plan. He was at pains to highlight "the difference betweeen community planning and the ordinary method of city-extension and suburb-building":

> *The normative idea of the garden-city and the garden-village is the corrective for the flatulent and inorganic conception of city-development that we labor with, and under, today. So far from being a strange importation from Europe, the garden-city is nothing more or less than a sophisticated recovery of a form that we once enjoyed on our Atlantic seabord, and lost through our sudden and almost uncontrollable access of natural resources and people.*

> *Until our communities are ready to undertake the sort of community planning that leads to garden-cities, it will be empty eloquence to talk about the future of American architecture.*[11]

Given such a stance in 1924, it should come as no surprise that Mumford in later life would resist not only Bucky Fuller's more grandiose designs, but other mega-projects like Paolo Soleri's 'sculpted' city-scape at Arcosanti. No one man's perspective could, in Mumford's view, ever be adequate to encompass the eventual flowering of living communities – but one could attend to the seed-bed, and prepare the ground. Mumford stood for local knowledge, for the traditional wisdom of the place, for the organic growth of cities from the inside out – and against grand 'plans' imposed from the outside (unless they were for 'garden-cities,' of course) or top-down by architects. While the mariner Bucky Fuller gazed over the boundless sea to the horizon, the town-and-country scholar Mumford stood firm upon the craggy land:

> *Out of the interaction of the folk and their place, through the work, the simple life of the community develops. At the same time, each of these elements carries with it its specific spiritual heritage. The people have their customs and manners and morals and laws; or as we might say more briefly, their institutions; the work has its technology, its craft-experience, from the simple lore of peasant and breeder to the complicated formulæ of the modern chemists and metallurgists; while the deeper perception of the 'place,' through the analysis of the falling stone, the rising sun, the running water, the decomposing vegetation and the living animal gives rise to the tradition of 'learning' and science.*[12]

Here, if anywhere, were the human roots in land and in custom that Bucky's grand design conceptions not only lacked, but intentionally cast aside. From another angle, though, Fuller and Mumford's work really did share a common root. They both sought the convergence of organic forms and cultural forms – the very convergence which so galvanized Emerson, their mutual progenitor, in the midst of his Adirondack sojourn. In an early essay called "The Marriage of Museums," Mumford lauds a proposal to build a connecting walk through Central Park to tie together New York's Museum of Art and its Museum of Natural History: "The physical connection will serve to emphasize a cultural borrowing which has at once introduced the presentiments of graphic art into the nature museums, and the organic conception of life into the art museum."[13] Where the two men differed was in their conception of just how such a fusion of Nature and Culture was to take place. Mumford would accept 'development' only according to the word's etymology: the unfolding of latent possibilities, growth from the inside out. Fuller understood the process in two steps: first, observing and abstracting from Nature the principles with which she works, and second, turning round and imposing these forms upon human artifacts and human dwelling; development from the outside. In *Technics and Civilization* (1934),[14] Mumford hailed Fuller's plans for the Dwelling Machine and the Dymaxion Transport as intriguing prospects worth watching. By the 1960s, however, in his devastating critique of *The Myth of the Machine* (1967/1970),[15] he had soured on a society which had opted for mechanism rather than organic life, weaponry over the arts of living, giantism rather than scaled invention, and a global economy rather than regional self-sufficiency. He took a famous passage of Fuller's – from one angle full of ironic humor and intended only to open people's eyes to the design integrity of Nature's own handiwork – as an occasion to mock what he saw as Bucky's under-dimensioned appreciation of human nature:

> *Man, observes Fuller, is 'a self-balancing, 28-jointed adaptor-based biped, an electro-chemical reduction plant, integral with the segregated stowages of special energy extracts in storage batteries, for subsequent actuation of thousands of hydraulic and pneumatic pumps, with motors attached; 62,000 miles of capillaries, millions of warning-signal, railroad, and conveyor systems; crushers and cranes … and a universally distributed telephone service needing no service for 70 years if well managed; the whole, extraordinarily complex mechanism guided with exquisite precision from a turret in which are located telescopic and microscopic self-registering and recording range finders, a spectroscope, etcetera.'*
>
> *Fuller's parallels are neat; the metaphor is superficially precise, if one discounts the airy, pseudo-exact statistical guesses. Only one thing is lacking in this detailed list of mechanical abstractions – the slightest hint, apart from his measurable physical components, of the nature of man.*[16]

Ouch! In this period, Mumford went on to disparage Fuller as a 'genius' (Fuller denied it) who wouldn't learn from his mistakes (hardly true), and leapt upon the Expo Dome fire as if it were Bucky's fault (it was not; a spot welder tried to finish up his repairs even though his fire extinguisher was empty). In some respects though, it is fair to say, Mumford's critique hit directly and uncomfortably on target. Fuller's 1943 plan for the rapid industrialization of Brazil, for instance, would have trashed the place far faster than the haphazard 'development' which currently acounts for an annual loss of 10,000 acres of rainforest; he actually advocated, for example, *bombing* hundreds of clearings in the forest and parachuting in machine tools to build timber mills and other industries, airports, etc. (He had of course no idea people were actually living there, but from Mumford's perspective such ignorance of the region on the part of 'planners' and 'developers' is precisely the problem.) Later in life, Bucky explicitly dismissed Mumford's concerns as 'aesthetics.'[17] "Don't worry," he tells a group of architecture students in Robert Snyder's film, "about making your work beautiful" –

> *You don't have to worry about 'beautiful' or 'pretty,' because if you have really understood your problem, if you solve it correctly, so life really goes on; if you do it so economically it is realizable: then it always comes out beautiful.*
>
> *That's why a rose is beautiful: as one part of the great regenerative process whereby the a priori design of the Universe is working. If you want to be part of that, you can't miss beauty.*

It was clear by this time that the two thinkers, both of whom had high public profiles (Mumford was President of the American Academy of Arts and Letters), had stopped listening to one another. By the time of his *Findings and Keepings* (1975), Mumford dismissed Fuller as "that interminable tape-recorder of 'salvation by technology,' " and lumped him with Marshall McLuhan, Daniel Bell and Arthur Clarke – as one of "those giant minds whose private dreams all too quickly turned into public nightmares."[18]

It is somewhat invidious simply to contrast the two thinkers. Mumford matured early, Fuller rather late; when Fuller began his ascendancy on the world stage during the 1960s, Mumford was already at his peak, and his powers were to decline drastically before his death. Mumford was from the outset a writer of mastery and grace; Fuller, for all his inventive neologisms, a labored and clumsy writer. Mumford had a lot to say about architecture, but never built or even designed anything of consequence; Fuller had little patience for drawing-room theories, but put most of his formidable creative energies into actually making things (including books, of course, which might be counted among his most enduring artifacts).

In later life, they rarely agreed about anything. If Fuller extolled alloys and metallurgy as proof that humans evolved by learning how to do 'more and more with less and less,' Mumford excoriated mining and metal-working as the rape of the Earth and the first instance of wage-slavery for the hapless human digging tools forced to work in the mines. If Fuller expected NASA's space program to provide a 'black box' kit for human dwelling here on Earth, Mumford saw the astronaut as "Encapsulated Man":

> *... a scaly creature, more like an oversized ant than a primate – certainly not a naked god ... a faceless ambulatory mummy ... coordinating his reactions with the mechanical and electronic apparatus upon which his survival depends ... whose existence from birth to death would be conditioned by the megamachine, and made to conform, as in a space capsule, to the minimal functional requirements by an equally minimal environment – all under remote control.*[19]

That remote control would of course be provided by the computers Bucky hoped might help co-ordinate world-around information and planning for his Geoscope. Mumford found computers an excuse for not thinking, just one more abdication of human responsibilities to the 'myth' that machines could bring about the human millennium. In puncturing all such futurist fantasies, Fuller's easily included, Mumford was taking aim at:

> *... the state that the mass of mankind is fast approaching in actual life, without realizing how pathological it is to be cut off from their own resources for living and to feel no tie with the outer world unless they are connected with the Power Complex and constantly receive information, direction, stimulation, and sedation from a central external source, via radio, discs, and television, with the minimal opportunity for reciprocal face-to-face contact.*[20]

Fuller and Mumford were both 'public intellectuals,' a vanishing breed on the American scene these days. Both were also astonishingly comprehensive autodidacts, another species nearly as extinct as the moa. Mumford was the more scrupulous and exacting as a scholar, and the more balanced as a writer. Fuller was the more inventive, always able to pull special effects – visual, tactile, geometric, and architectural surprises – out of his bag of tricks. It is important to recognize that although both eventually lectured at colleges and universities, neither held advanced, 'specialized' academic degrees. Their generation was happily less hamstrung than our own by 'experts' claiming dominion over every conceivable petty 'field' of study. Both took justifiable pride in their role as self-taught generalists, and both railed against the 'barbarism of specialization.' With obvious relish, Fuller used to relate the following story about academic myopia:

In New York City a few years ago, two learned societies met in separate hotels on the same day to discuss strikingly similar topics. The one group consisted of biologists attempting to determine the major factors causing extinction in species of flora and fauna. The other study group consisted of cultural anthropologists grappling with the causes of extinction among human tribes and groups. In both cases, the primary factor leading to extinction turned out to be the same: *overspecialization*. Ironically, neither group of 'specialists' had the least inkling of the other group's findings. The moral of the story? If we're not careful, we may very well specialize ourselves right out of existence.

Yet Fuller and Mumford were not unlike those two scientific societies in this respect, that neither could really see what the other saw. In his quest for the Whole, Fuller came to favor the simplicity of a heteronomic viewpoint: the whole holding sway over all its parts, coordinating every element to play its proper role. For his part, Mumford came to emphasize complexity and autonomy: each part a free whole in its own right, trusting the larger Whole to take care of itself. Of course the struggle between these two viewpoints – the One and the Many, *'en kai polla* – has been the outstanding Western intellectual quandary ever since Plato opted for the One, and Aristotle for the Many. By the 1960s, the two thinkers were simply talking past each other. Each had a comprehensive view which took into account nearly everything – except of course the annoying existence of the other, perhaps equally plausible, viewpoint. Now these two thinkers knew about one another, read one another, and for 50 years scrupulously avoided any but the most oblique dialogue with one another. Why?

It would be superficial merely to label Bucky a technological optimist, and Mumford a pessimist – although the caricature readily fits some of their isolated pronouncements. Mumford tended to speak of what was and what had been, and Fuller of what might be; but Mumford also wrote eloquently of the future in "The Drama of Renewal"[21] and "The Flowering of Plants and Men,"[22] while Fuller could be withering in his scorn for the stupidity and wastefulness of status quo science and technological folly. (His satire on the fictitious corporation "Obnoxico" might well describe the attitudes of many a multinational colossus from the 1990s to the present-day).[23] Both were prophets of the ecological era long before most people even knew what the word meant. But what do prophets do? They *announce* (there's Fuller, looking forward), or they *denounce* (there's Mumford, looking warily back over his shoulder, although he professed the role went against his grain). And there they leave us, these two prophets, with the entire twentieth century as the ambivalent proving ground for their theories. It is, therefore, not a little difficult for those of us who follow them to discover exactly *which* 'holistic' vision is the genuine article, which the more real insight. In a certain sense, they leave us stranded between past and future, memory and imagination, origins and destinations.

In the present, they both remain fairly influential mavericks. Mumford, lamentably, is not required reading in most college curricula nowadays, even for the liberal arts. His scope is too sweeping, the questions he asks too challenging to be compatible with the technical 'training' of worker-drones favored by today's educational 'marketplace.' And Fuller offered an alternative science, an exploratory strategy for independent scientific and technological inquiry, not the sort of thing sought by young people trying to fit into the job slots and training programs offered by big corporations. Of course since the discovery of 'fullerenes,' Bucky has become trendy again; a revival of interest in his structures may yet lead people in the sciences back to his books and ideas. On the other hand, the trenchancy and effectiveness of the critiques of technocracy launched by Mumford and others influenced by him (Jacques Ellul, for instance, or Ivan Illich) has meant that Fuller's work and its import has been almost universally neglected by serious scholars in the humanities.

Mumford stands for the sensibility that has won over many of today's most articulate scholars of technology and human values. It is, with various nuances, an anti-technology, anti-development sensibility, critical of globalizing trends, respecting local knowledges and traditional customs, suspicious of planning and planners, mustering defenses for threatened ecologies and endangered cultures alike. It is this very sensibility that leads one to take seriously the 'archaic' insights of native peoples (as we tried to do with respect to the Mohawk Nation earlier on), as well as the integral relationship of land and landforms – bio-regions – to human culture, dwelling, thinking, spirituality, etc. An easy familiarity with Mumford's themes has long been perceptible in the programs of the German Green Party, for instance (even as the German auto industry reconsiders Bucky's three-wheeler Dymaxion Transport for mass-production). Today we are, or ought to be, much more critical of universal 'development' schemes, whether Fuller's or anybody else's; we rightly question whether everybody on the planet needs or even wants to 'develop' in the same way, or toward the same Western goals.

For his part, Bucky not only made palpable the possibility of an alternative science (that's really what *Synergetics* is all about), but he showed off some splendidly creative examples of what might be achieved by somebody taking roads less traveled than the current destructive path of industrial monotechnics. By contrast, what many humanist thinkers nowadays are really looking for – and even finding, in indigenous cultures for example – are alternatives *to* modern science and technology, so that Bucky tends to get lumped in with all the rest as just another technocrat. For their part, the scientists, so very well trained in their respective specializations, are just not ready or willing to put up with all the paradigm shifts Fuller requires of them. It's still business-as-usual, not 'utopia or oblivion.' Bits and pieces of Bucky's vision are indeed taken up – a word about synergetics for the mathematicians (I've heard it called 'trivial,' in Silicon Valley of all places); a page or two on geodesic domes for the architecture students; a footnote on oc-tet trusses for the budding engineers; a sidebar on the Dymaxion Map for the cartographers. But almost nowhere in North America, except perhaps for a brief time at the Buckminster Fuller Institute in California, do scholars and students get a chance to encounter the whole of Fuller's thought, let alone to challenge it by the equally comprehensive, equally 'holistic' thought of Lewis Mumford. Which is not only a shame, but a shameful failure to follow up on the fundamental questions of technology and human values both writers raised in their own ways. Recent developments, however, may soon make scholarly collaboration a little more practicable. In 1999, Bucky's 'Chronofile' – 1400 linear feet of file boxes, the largest collection of papers from any single human being in the twentieth century (he kept everything!) – was relocated from the Fuller Institute to Stanford University, where it has now been catalogued for easier access.

* * *

We won't be able to get Fuller and Mumford together for the knock-down, drag-out conversation they should have had, but we can see why it is imperative to continue their dialogue as best we can *in absentia*.[24] For one thing, if you never think about their divergences, you never have a chance to see where they do, surprisingly, converge. Fuller's was a cosmological vision; he looked mainly to the physical Universe and to Nature, and discerned there the holistic dimension of cooperation – *synergy* – between all her elements and principles, which he then extrapolated (sometimes perhaps a little insensitively) to human affairs. Mumford's was a humanist's vision; he looked mainly at the deeds and misdeeds of human societies and discerned there the holistic dimension of cooperation – *community* – swiftly being eroded by megatechnic schemes which ignore both cultural context and human scale.

Today the same split can divide 'deep' ecologists who see humans as one species in a living, interdependent cosmos, from campaigners for social justice who seek to liberate human beings from structures of oppression in all its social forms. The two visions are not incompatible, but the contexts in which they are spelled out tend to diverge sharply. The common thread is cooperation, though its exponents sometimes let this slip in their enthusiasm for their own particular angle. Both Fuller and Mumford thought they had bridged C. P. Snow's 'two cultures' of the sciences and the humanities, but the bridge seemed to collapse where one another's viewpoints were concerned. Now, half a century on, can we meaningfully coordinate the cosmological/scientific and the anthropological/humanistic visions? Let's see if we can't find some common ground …

Both Fuller and Mumford thought that modern technological culture had taken a wrong turn in the twentieth century. Two World Wars pretty well clinched the case. Both sought organic patterns and natural principles on which to base the turn (Fuller), or the return (Mumford), to a more humane culture. Most alarmingly, both thinkers found the present direction of world's nations and industries suicidal, for equally firm ecological and social reasons. But technology meant something different to each of them. For Mumford, technology was mechanism and, though a culture might benefit from some degree of mechanistic constraint (ritual tabus, for example), the total submission of organism to mechanism was lethal. Fuller, on the other hand, considered technology to be what Nature already does so elegantly – the hydraulic leveraging that holds up the branches of a tree, the tensile gravitation that keeps our planet in its orbit – and which we humans mimic only clumsily and mechanistically. In Fuller's view, we suffer from our ignorance of Nature's supremely efficient technologies, attention to which would greatly refine our own artifacts and enhance our lives. In Mumford's, we have let our technology come not only between us and Nature, but also at the cost of our own human nature.

The distinction is important, because it reveals an unexpected concurrence. It is grounded in the deep respect both thinkers share for the mystery at the deepest core of life's processes. Mumford put it this way: "The ultimate gift of conscious life is a sense of the mystery that encompasses it."[25] And in a similar vein, reminding us of the proximity of spirituality to the dwelling place, and of cultural to physical space: "Religion concerns itself with the reaction of Man in his wholeness to the Whole that embraces him."[26] As if in chorus, Fuller's "Intuition – A Metaphysical Mosaic" opens similarly:

Life's original event
and the game of life's
order of play
are involuntarily initiated
and inherently subject to modification
by the a priori mystery,
within which consciousness first formulates
and from which enveloping and permeating mystery

consciousness never completely separates,
but which it often ignores
then forgets altogether
or deliberately disdains.[27]

We may simply have to concede that this concurrence of Mumford and Fuller on such a basic theme might be part and parcel of that very mystery. It's as if some inexpressible harmony, or at least an unseen complementarity, lay behind their passionate refusal even to consider one another's perspectives. The whole of their relationship, if you think about it long enough, comes to a good deal more than the sum of its parts taken separately. For the sake of balance, we shall hand over the last word of this chapter to Mumford. Had he taken the following exploration of triangles and hexagons just a bit further, he'd have run smack into Fuller's isotropic vector matrix (the very source of the synergetic geometry), which derives directly from the configuration Mumford so lovingly describes here:

Suppose we were all equilateral triangles. Accepting our basic triangularity, we could not conceive of any change, except that of becoming more or less of a triangle. We might think of becoming a bigger triangle or a smaller one, turning from an isoceles triangle to a scalene triangle, almost to the point of being flattened out; but the one thing that would seem indisputable to us would be that we could not keep our precious triangularity if we tried to cut any other kind of figure. Yet that would be a delusion; and it would be the kind of delusion that is brought about by a failure to conceive the role of love.

Let us endow two right triangles, absolutely equal in every respect, with the power to fall in love: that is, to delight in the constant presence of the other, and spurred by the desire to meet and mingle on the closest possible terms. The nearest that a triangle could get to connubial bliss with another triangle would be if 'he' superimposed 'himself' on it, and in order that our identical triangles should not merge their identity completely, we will suppose that the apex of one triangle in the ideal state of their union would intersect the base of the other. At that moment a being unknown in the world of triangles will come into existence: a star. A quite remarkable star, with six points, and with an internal figure in a central position, holding the parts together: a hexagon. What is more, at the tips we will find that this union has begotten six little triangles, just like their father and their mother, only smaller. Mark that the lines and angles of the original triangles remain unchanged: yet in the combination they show new properties, unknown in a purely triangular world. In this scenario none of these little triangles would have existed if the big triangles had not 'fallen in love' and sought to merge their identities.

This parable not merely sums up the nature of emergent change, using existing components to make a radically different pattern or 'Gestalt': it also demonstrates, by the simplest of abstractions, the unexpectedly creative interaction that takes place in loving association! That association had its beginning at the very moment when the sexual differentiation into male and female began in the plant world, and was carried to a climax in the ostentatious sexuality of the flowering plants, long before vertebrates emerged.[28]

27 August 1956

III. Reflections

Anima Mundi

Mumford's amorous right triangles (90°/45°) combine to give us this figure:

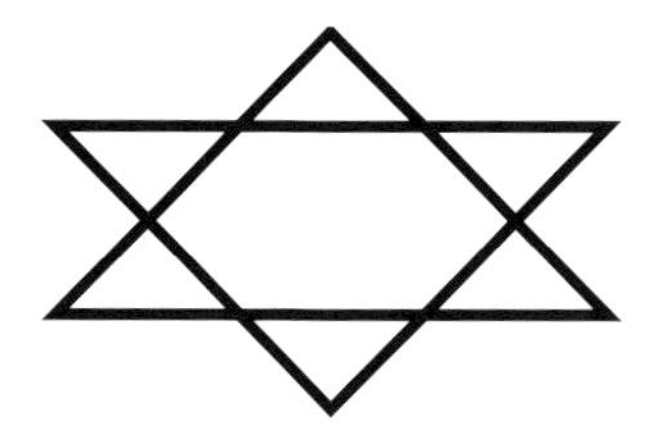

And there ends his reflection upon "the nature of emergent change," i.e., the synergy of any new "Gestalt." Had he stayed instead with the equilateral triangles (60°) he supposed we all were at the start, the figure would have shifted slightly to look like this:

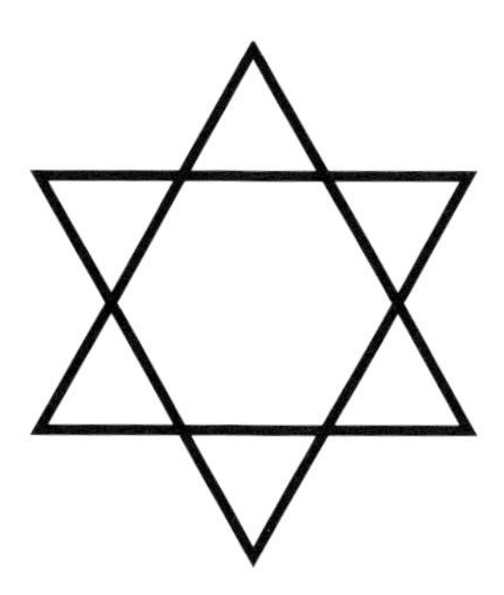

Now if you look at this figure long enough, and play around a bit with its properties, the reflection of the ascending and descending triangles, along with the generation of attendant stars and hexagons, could go on practically forever. If you begin with equilateral triangles, and articulate these into tetrahedra, you find octahedral voids popping out between them to form the 'oc-tet truss,' the strongest load-bearing structure known to engineering. Continuing the procedure to link up all the vertices discloses an X-ray of Archimedes' cuboctahedron, a figure Fuller first dubbed 'The Dymaxion' and later Vector Equilibrium, a pattern in which the tensile and compressive forces are exactly in balance. You have now arrived at the simplest form of the *isotropic* ('same everywhere') *vector matrix* which Fuller called 'Nature's Coordinate System.'

Just as points of reference, I shall have to make short work of the many scientific corroborations now available for Fuller's intuition that Nature is using this matrix,

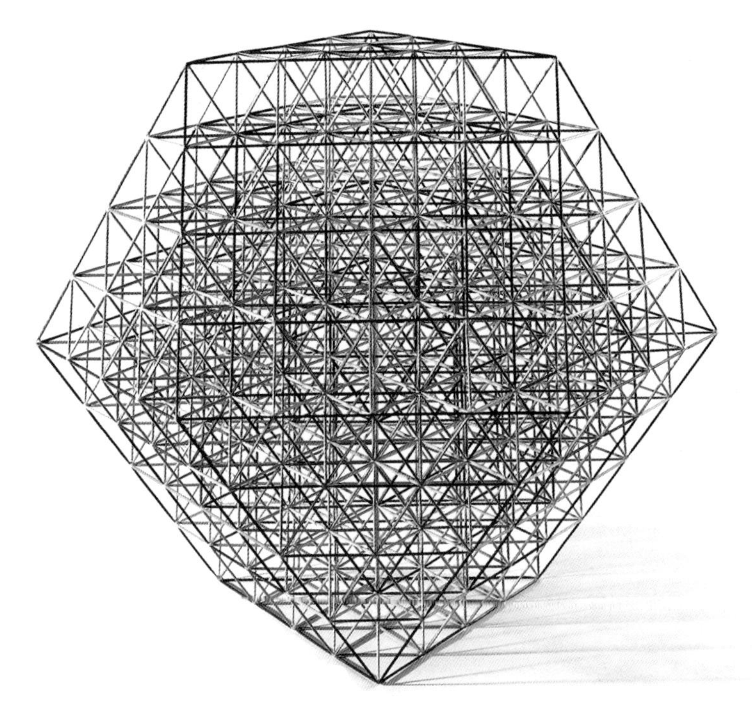

3D Isotropic Vector Matrix (Estate of R. Buckminster Fuller.)

with its 60° coordination in every direction, for her most basic atomic, molecular, organic and astrophysical configurations.

A few highlights:

- In 1863, Nicholas Van't Hoff discovered that the carbon molecule, and therefore all organic carbon-based chemistry, is tetrahedrally co-ordinate.
- In 1958, Linus Pauling demonstrated by diffraction grating analysis that all metallic (i.e., 'non-organic') structuring is also coordinated tetrahedrally.
- In 1975, Harvard mathematician Arthur Loeb contributed an "Afterpiece" to Fuller's *Synergetics* in which he demonstrated that crystal structures of all sorts were accommodated as well by Fuller's isotropic vector matrix as by conventional cubical coordinates. (In point of fact, the cube does appear in Fuller's matrix, but only as a sort of optical illusion if you're looking at its octahedra and tetrahedra).
- The geodesic domes which Fuller generated from variations on this matrix turned out to be identical in structure to various viruses and antibody cells, which is probably a clue as to why they're so tough.
- The mathematical projections of the strong binding forces of atomic nuclei, which quantum physics terms mesons and baryons, turn out to be vector equilibria and tetrahedra, respectively. Bucky was peering into the very heart of matter, where he kept running into old 'friends' (figuratively speaking) from his earlier geometric explorations.
- In 1985 Carbon 60, a remarkably stable cluster of 60 carbon atoms said to be the form of carbon carried by diffuse interstellar gas clouds (and thus the probable

seed-form of carbon-based chemistry in the Universe at large), was synthesized in the lab and called 'Buckminsterfullerene' in honor of the man whose name had become nearly synonymous with icosahedral spheroids. Today there is an entire family of 'Buckyballs,' variant 'fullerenes' custom-designed for diverse purposes.

- The isotropic vector matrix corresponds to the interference pattern formed by the propagation of electromagnetic energy in a vacuum – Einstein's universal constant, the 'c^2' of $E = mc^2$ – and therefore brings into a single image the sunlight that animates life on Planet Earth and all the new media of electronic, digital and fiber optic communications technology. It is the pattern language – the media matrix, I have called it elsewhere[1] – of a Universe in radiant communication with itself in all directions.

Indeed, I would go so far as to declare this 'omniintertriangulated' matrix to be the very structure of *illumination,* in every sense of the word. At any rate, there is plenty of empirical corroboration from the sciences that Fuller was indeed on to 'Nature's Coordinate System' in a profound way. Following Einstein's proofs for a curved, finite Universe, Fuller banished the bugaboo of infinity from both his geometry and his conception of Cosmos, yet the ramifications of his 'whole systems' approach seem to be very nearly endless. He basically contends that Nature can be so prolifically diverse in reproduction and variation only because she is always most economical in *structure.* The isotropic vector matrix can also be directly derived from the closest-packing of spheres; it models 'least effort' configurations and structural strategies – unlike the cube, which distorts and wastes space, time, and energy prodigiously. At the center of this matrix is the Vector Equilibrium (cuboctahedron), the shape Fuller sees as the unfathomable 'zero' of the energetic-synergetic geometry. Indeed, Krause and Lichtenstein consider his 'Jitterbug' Transformation of the Vector Equilibrium – a fairly simple wooden stick VE model with latex rubber joints – to be Fuller's single most important invention:

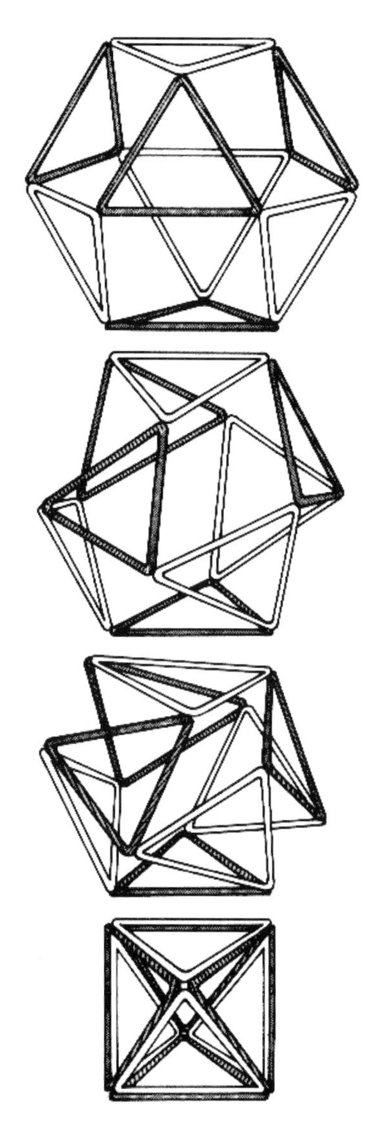

The Jitterbug Transformation
(R. B. Fuller, Synergetics, *NY, 1975)*

> *In the Jitterbug Transformation, a cuboctahedron is placed over an icosahedron to form an octahedron (which can be folded into a triangle). For two thousand years, the Platonic bodies stood stood statically and proudly next to one another, and then Fuller ... used ... his energetic-synergetic geometry to show that they were closely related in structure, and that they could be understood as the phase transitions*

of a transformative 'loop,' specifically, a periodic swinging back and forth between two 'end' phases.[2]

Since Fuller saw not only into but *through* this shape in such a remarkably thorough way, we should allow him to speak for himself about the Vector Equilibrium and its structural implications:

The vector equilibrium is the anywhere, anywhen, eternally regenerative, event inceptioning and evolutionary accommodation and will never be seen by man in any physical experience. Yet it is the frame of evolvement. It is not in rotation. It is sizeless and timeless. We have its mathematics, which deals discretely with the chordal lengths. The radial vectors and circumferential vectors are the same size.

The vector equilibrium is a condition at which nature never allows herself to tarry. The vector equilibrium itself is never found exactly symmetrically in nature's crystallography. Ever pulsive and impulsive, nature never pauses her cycling at equilibrium: she refuses to get caught irrecoverably at the zero phase of energy. She always closes her transformative cycles at the maximum positive or negative asymmetry stages. See the delicate crystal asymmetry in nature. We have vector equilibriums mildly distorted to asymmetry limits as nature pulsates positively and negatively in respect to equilibrium. Everything that we know as reality has to be either a positive or a negative aspect of the omnipulsative physical Universe.

... As the circumferentially united and finite great-circle chord vectors of the vector equilibrium cohere the radial vectors, so also does the metaphysical cohere the physical.[3]

If you were a scientist, these astonishing correlations of the isotropic vector matrix with empirical evidence from various realms of the physical world might be enough to keep you busy for a lifetime, as they did Bucky. But a more artistic or philosophical or religious sensibility might see further reflections, more deeply buried perhaps, but in some ways even more extraordinary still. Many would immediately recognize here the Jewish Star of David or, with interlacing dark and light triangles which tradition sees as the union of body and soul, the Seal of Solomon. As if to reinforce the connection, Canada's premier stained glass pioneer Eric Wesselow produced his 'growing' *Magen David* (Star of David, p. 79) after studying Fuller's vector equilibrium.

If you had instead a classical bent, you might very well see hidden here the Pythagorean *tetraktys:*

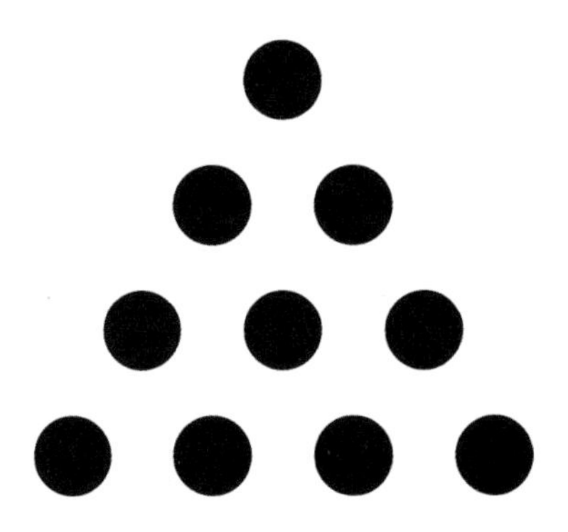

Wesselow, The Growing Magen David (1990)

From this angle, the vector equilibrium at the heart of Bucky's matrix is not just a stereoscopic version of Archimedes' cuboctahedron, but a tetraktys raised to the second power: when one tetraktys is inverted and overlaid upon another, the central ten dots coincide, adding only an inverse outer triangle; the whole forming the familiar 12-around-1 vector equilibrium configuration of closest-packed spheres:

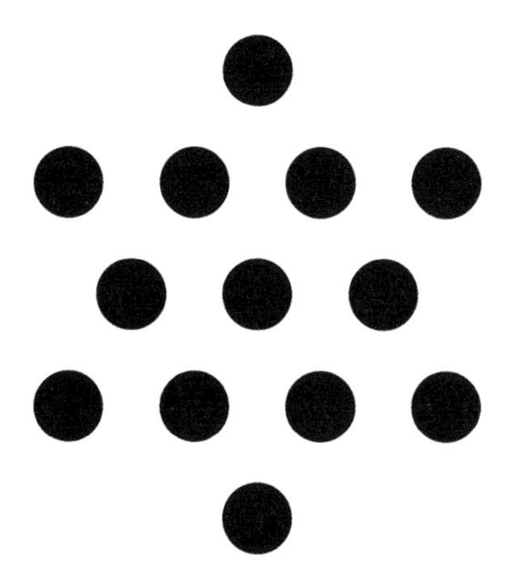

Tilt this figure 30° over to one side, and you are again face-to-face with Fuller's 'Dymaxion,' the lines of force connecting closest-packed spheres:

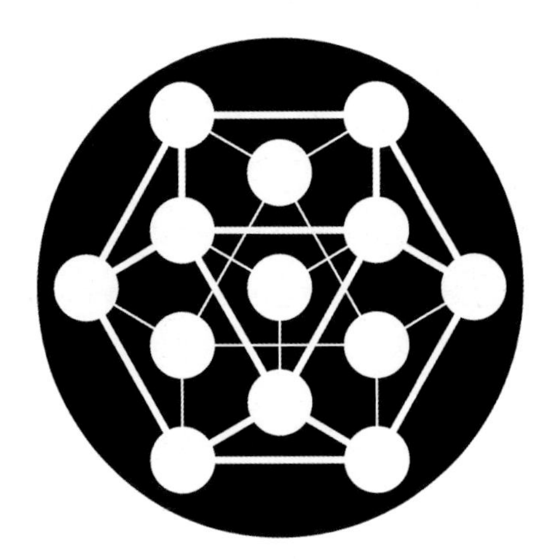

And here you may recall that Pythagorean initiates were required to swear their allegiance (as well as their promise not to reveal the secrets of the Order) by the Oath of the Tetraktys: "I swear by Him who reveals Himself to our minds in the Tetraktys, which contains the Source and Roots of everlasting Nature."[4] And what was the celebrated doctrine of the Pythagoreans – the Golden Mean, moderation in all things – if not this balance and this equity played out in the human and social world? You would also be reminded very strongly of Plato's effort to describe the cosmogony, the genesis of all things, with an arresting geometrical image probably as old as human memory: *the soul of the world.* We spoke earlier of Emerson's sense of it. Here we find Socrates listening to the explanation offered by the Pythagorean Timaeus, drawing from sources already ancient in their own time (c. 500 BC):

> *We may say that the world came into being – a living creature truly endowed with soul and intelligence by the providence of God … Wherefore he made the world in the form of a globe, round as from a lathe, having its extremes in every direction equidistant from the center.*[5] (*Tim.* 30c, 33b)

And the creator proceeded to divide this globe, Timaeus recounts, by triangulating it in every direction following the harmonic intervals of the tetraktys. The whole was then reconstituted into an interlocking system of perfect circles within circles, assumed to correspond to the orbits of the seven known planets. The permanent elements of this world are described as the five regular polyhedra – the famous Platonic 'solids' – said to govern all form and recurrence in the natural world. In this image, we catch sight of the fusion of two traditions: the ancient Mediterranean Great Mother Goddess (*mater, matrix:* womb), the maternal 'sphere' which gives birth, which nourishes and sustains, as well as the more abstract mathematical 'key' to her prolific diversity, the tetraktys, which also describes the harmonics of the lyre's vibrating string.[6] Even as the soul of the world is fused with its material body, we are encouraged not to lose sight of 'her' originally spherical shape:

"God geometrizes", as Plutarch described Plato's vision in the Timaeus. *(Minature from Genesis page, mid-13th Century French Bible, Codex 2554, folio IV, Osterreichische Nationalbibliotetek, Vienna)*

> *Now when the creator had framed the [world] soul according to his will, he formed within her the corporeal universe, and brought the two together and united them center to center. The soul, interfused everywhere from the center to the circumference of heaven, of which also she is the external envelopment, herself turning in herself, began a divine beginning of never-ceasing and rational life enduring throughout all time. The body of heaven is visible, but the soul is invisible and partakes of reason and harmony, and, being made by the best of intellectual and everlasting natures, is the best of things created.* (*Tim.*, 36e)[7]

From Plato's spherical vision of perfect circles and regular 'solids,' there is a nearly unbroken tradition (if we forgive Euclid his excess of analytic zeal in flattening the whole thing out) which culminates in the 'spheres within spheres' of the medieval Neoplatonism to which Dante, among many illustrious others, was heir. The immensely influential *Celestial Hierarchy* of the Pseudo-Dionysius (c. 550 AD), who has quite rightly been called 'the fountainhead' of Western mysticism, set the pattern of nine metaphysical spheres 'animated' by their respective intelligences.[8]

A clarifying note: The 1990s witnessed a remarkable revival of popular interest in angels, but these are not of that sort. Supernatural guides, angels (good or bad) on your shoulder, personalistic 'guardian' angels and so forth most likely derive from the Jewish tradition of the two *yetser,* the twin urges of the human heart,[9] often clouded by an atmosphere and an iconography derived from the dualistic Zoroastrian tradition: the powers of darkness versus the legions of light.[10] The impersonal Neoplatonic tradition of angelic intelligences moving the celestial spheres is something else again, and much closer to Fuller's notion of primordial, 'timeless, metaphysical' patterns animating the transactions of energy in the atomic, molecular, and biological domains. As Ezra Pound observed: "The statements of analytic geometry are 'lords' over fact. They are the thrones and dominations that rule over form and recurrence."[11] Matthew Fox and Rupert Sheldrake have recently attempted in *The Physics of Angels* to recoup some of these cosmological functions of angels as the 'intelligences' governing various domains of the real, but they lack the rigorously disciplined trinitarian imaginations of the Pseudo-Dionysius, Scotus Eriugena, Bonaventure, or Dante, to mention only some of the tradition's most illustrious exemplars, and tend to get tangled up in wishful fancies all their own.[12]

The same cosmology also captivated many of the great medieval figures of Islam, including Avicenna (Ibn-Sina) in his *Tale of Hay Ibn-Yaqzan* and Ibn el-Arabi, specifically in his *Mohammed's Ascent to Paradise,* which may have provided a structural template for Dante's *Commedia.* Speculations like these had underpinned the structural insights of the Masons and helped spark the extraordinary burst of creativity that built the cathedrals throughout Europe during the Middle Ages. Interestingly for our purposes, the Gothic Master Builder Diagram worked on the same icosahedral pattern, incorporating the *phi* ratio and other sacred proportions, which Bucky used to build most of his domes.

By contrast, the Byzantine cupola which signals an architectural continuum between ancient Constantinople's Hagia Sofia and St. Mark's in Venice, and is also used to good effect in the twentieth-century Shrine of the Immaculate Conception in Washington, D.C., employs the four triangular 'pendentives' from the vector equilibrium to support a round dome on a square base – a configuration which

Angels & Angles

I may well be falling into the same trap as so many 'speculative' thinkers before me, but it seems to me fairly straightforward to translate Fuller's 'geometry of thinking' into the celestial spheres of the Pseudo-Dionysius. We need only move from the simplest such spherical forms through to the most complex.

Beginning with the spherical triangle and spherical tetrahedron, which unfold into the seven great-circle models outlined in our Synergetics Primer (Unfolding Wholes, Appendix A), we may re-assemble the nine 'intelligences' of the angelic hierarchy into three tiers of three, mirroring and moving outward from their unfathomable divine source in the Trinity:

The first tier: Spherical triangle (Seraphim/heat), spherical tetrahedron (Cherubim/light), spherical octahedron (Thrones/matter).

The second tier: Spherical vector equilibrium (Dominations), spherical icosidodecahedron (Virtues), spherical rhombic dodecahedron (Powers).

The third tier: 12-great-circle Vector Equilibrium (Principalities), 10-great-circle icosidodecahedron (Archangels), 15-great-circle triacontahedron (Angels).

To play out the analogy, if there is a 'fallen' angel in this schema – fallen precisely *outside* the original matrix – it is the square frame and *the cube* (which the Renaissance turned into the basis of the one-point perspectival system and the rectilinear grid spawned from it), the chief mischief-maker now exercising unchallenged 'dominion' over all the world ... or at least over most of its 'profane' architecture.

Granting that such an arrangement turns archangels into Buckyballs, you may take or leave it as you see fit. I can only give you the correspondences as I see them. Fuller presents the physical evidence (see Appendix A); the Neoplatonic tradition reinforces the metaphysical coherence of the whole.

Not to make too much of an acoustic pun, but it might be said that angles and angels fit snugly together in this worldview from the very beginning.

This is a worldview where the direct experience of revelation takes precedence over logical inference, which in turn takes precedence over empirical data ... in other words, a sacred, hierarchical worldview more or less precisely opposite to our own modern, secular, techno-scientific reductionisms.

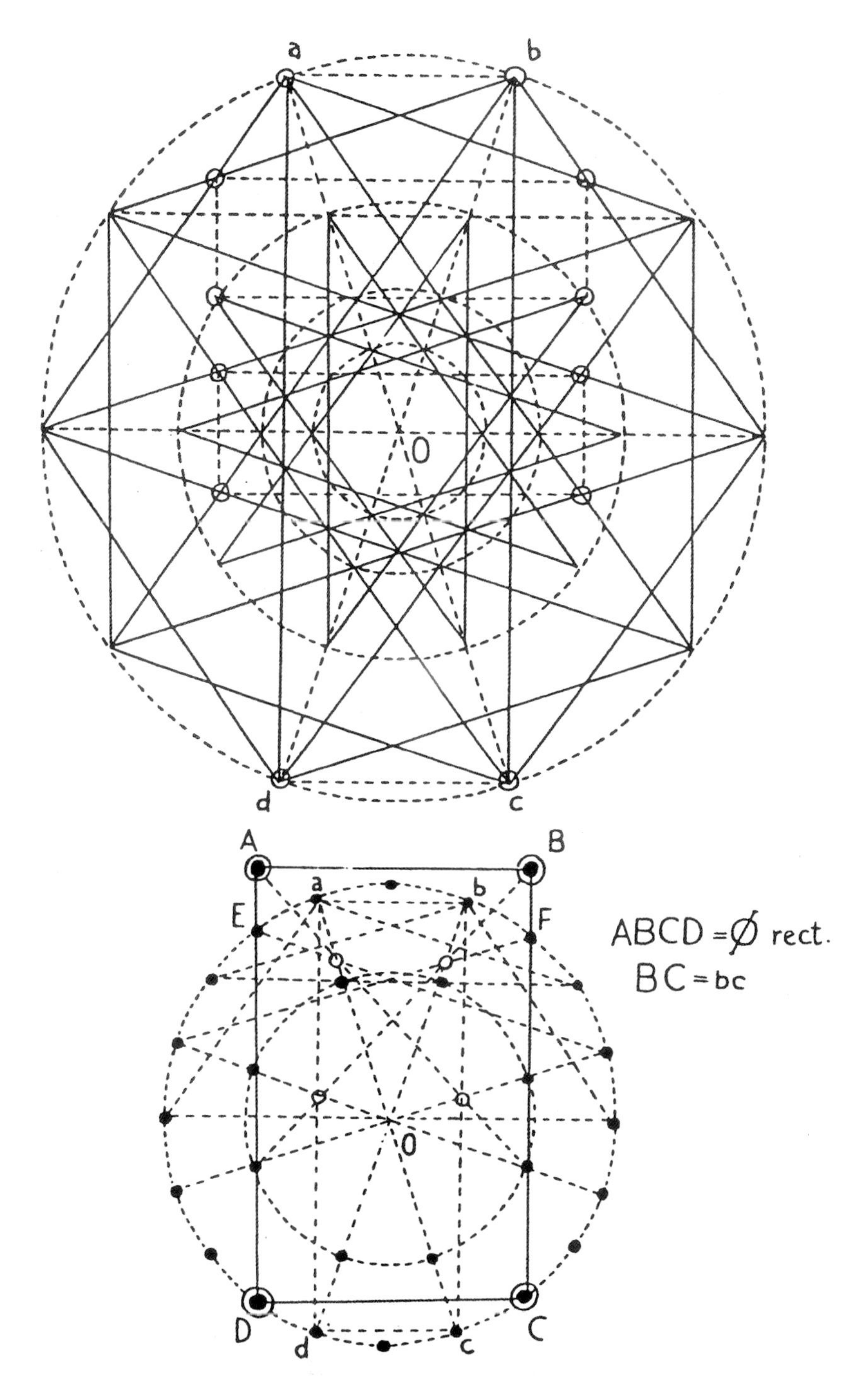

The Gothic Master Diagram.
(M. Ghyka, The Geometry of Art and Life, *New York, 1946)*

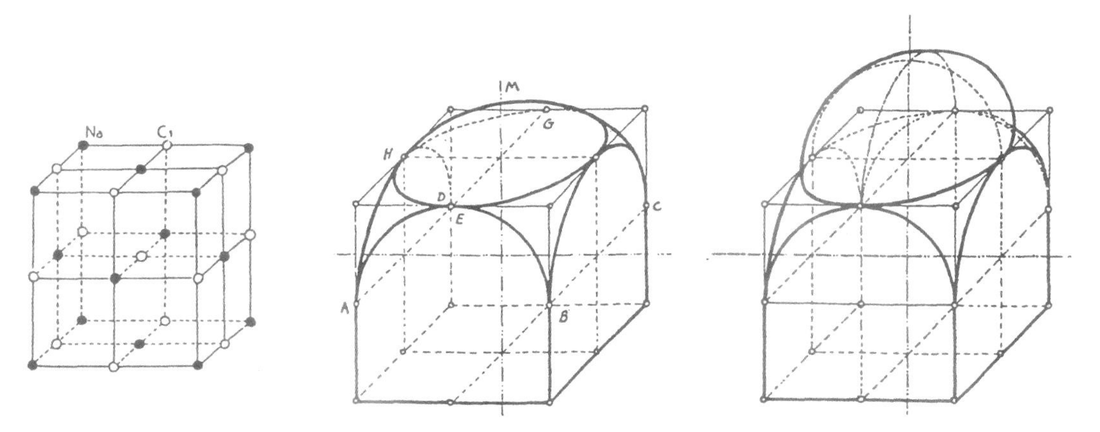

NaCl (salt), Cuboctahedron and Byzantine Cupola. (Ghyka, Geometry, *NY, 1946)*

turns out to be akin, oddly enough, to the molecular structure of common table salt.

Even as late as the Renaissance, Nicolas of Cusa would affirm in his *Of Learned Ignorance* (1440) that he shared with the Neoplatonists the insight that "the Soul of the World is an unfolding of the mind of God," which "necessarily co-exists with matter, from which it receives a limitation." Unlike Bucky, however, Nicolas permitted infinity into his geometry, thereby unduly muddying the waters for centuries. In his *De divina proportione* (1509), Luca Pacioli offered three-dimensional models of the Platonic solids, truncated and stellated the icosidodecahedra Bucky would later use to build his domes, and called on his close friend Leonardo da Vinci to render these complex figures for publication, probably from models.[13] Significantly for our purposes, Pacioli was the first geometer in the West to insist on making models rather than relying solely upon mathematical abstractions, a pragmatic desideratum Bucky shared with his own great teacher, H.S.M. Coxeter, Professor of Mathematics at the University of Toronto, to whom (along with "all the geometers of all time") *Synergetics* itself is dedicated.[14] In illustrating Pacioli's book from such models, Leonardo came up with a system for rendering these visually challenging polyhedra which has been employed in practically every subsequent treatise on geometry, perspective and astronomy. (These figures were, incidentally, the only set of Leonardo's designs published during his lifetime). In the early seventeenth century, Johannes Kepler was still trying to construct a precise geometrical model of the Platonic World Soul. But when his own heliocentric model of the solar system superceded the geocentric Ptolemaic astronomy, with which the *anima mundi* had been too closely (and quite arbitrarily) correlated, the brashly successful new astronomy cast into historical oblivion the very idea of a living and geometrically integrated World Soul, surely a classic case of throwing out the baby with the bathwater …

Stripped of its various geometrical elaborations, the *anima mundi* still retained in mainstream Christian theology, both Latin and Byzantine, the very basic intuition which we moderns are only lately coming to retrieve, namely that the Cosmos is *animate* – i.e., 'en-souled'; that it has its own life and integrity and contains its own principles of movement, growth and regeneration. It is even Biblical, as Wisdom 1:7 attests: *Spiritus Domini replevit orbis terrarum;* the Spirit of God fills the whole world, this terrestrial orb.

Star-Polyhedra after Leonardo. (Ghyka, Geometry, *NY, 1946)*

Augustine saw no reason to refute it, Origen considered it likely, and the idea is further refined by Thomas Aquinas and most of the other great Schoolmen. The Latin Church only condemned the notion as 'pantheistic' when the World Soul was *identified* with either God or the Holy Spirit, judging that such an identification would abridge not only the transcendence of God but the freedom of human beings, who would thus find themselves locked into a totally predetermined world.

In many ways, as noted earlier, the *anima mundi* is a notion most closely akin to the Great Mother Goddess of the Mediterranean basin, but Mother and Father, Earth and Sky, *Cosmos* and *Theos,* surely should not be conceived as antagonists. They are lovers, and their union in the sacred marriage *(hieros gamos)* forms a perennial theme in world religions.[15] Unfortunately, due in large part to the predominance of rationalistic, mechanistic and reductionistic methods in the early days of the natural sciences, it took only a couple of hundred years for modern Western society to forget altogether this vision common to ancient, classical, medieval and renaissance culture: a vision of the Universe as a *living* reality in which angelic intelligences move the celestial spheres, demons stalk the underworld, and spirits of all sorts populate the interstitial spaces. Even today, ecologists have constantly to remind us "that the universe is alive," as the Indian Brahmin Iarchus put it to the Pythagorean Apollonius of Tyana at about the time of Christ.[16] The crucial intuition is that the microcosm of the human soul mirrors the character of the whole Universe: "As above, so below."

But science has not stood still since it abandoned the all-inclusive *anima mundi* in its enthusiasm for concrete empirical data. Indeed, science today is not a monolithic edifice, but a constantly revised set of only-partially-overlapping theories, hypotheses, models and metaphors. I would submit that modern physical and biological science is slowly but surely rediscovering this primordial *matrix,* the ever-underlying *anima mundi.* Buckminster Fuller's 'geometry of Nature' will have to stand in as a signal case in point for this argument, larger and deeper in scope than we can encompass here. The vigorous 'renegade' cosmological theories of Fred Hoyle and his disciples come to mind as a good place to start ... One would have to include the vast 'recycling' universe of Hoyle's Quasi-Steady State cosmology (still awaiting observational verification), which resembles Bucky's "eternally self-regenerate scenario Universe" in important respects; one would have to outline (as John Gribbin does in *Stardust*)[17] the internal nuclear dynamics of stellar evolution, with all the emerging elements geometrically shaped by the constraints of the strong and weak binding forces of the atom; one might take quite seriously Wickramasinghe and Hoyle's revived 'panspermia' idea,[18] whereby life arrives on planet Earth from comets carrying freeze-dried bits of bacteria and viruses (with their 'geodome' structure) from interstellar dust; and finally one might arrive at all the forms of life on this planet through a synergetic process like Lynn Margulis's 'symbiogenesis,' by now well established,[19] which finds free-floating microbes and bacteria 'co-operating' to form eukaryotic cells and eventually multicellular life. Add these partial visions together and you begin to find yourself, as the ancients did, at home in a *living* Universe. Some of the details are still somewhat obscure, or speculative, or very technical, but the overall pattern directly challenges the mechanistic model that has dominated modern science from Newton to Dawkins. It is perhaps enough for us now to note that this momentous rediscovery of the *anima mundi* has gone almost entirely unnoticed, largely because work in the various fields involved has been in the main highly specialized and the vision of the whole can scarcely emerge in any one scientific discipline alone.

In this domain of cosmology, I must confess myself an unabashed Bucky-booster. By using only his own mind, deliberately unencumbered by the conventional Euclidean assumptions, he may well have seen more clearly into the very structure of matter than anyone before him or since. His achievement is remarkable, and almost unparalleled. Challenged in the 1970s by Ed Applewhite to lay out the entire energetic-synergetic geometry in comprehensible form, he undertook to produce in the *Synergetics* volumes the groundwork for an integrated alternative science based on the 60° coordination of the tetrahedron unfolding into the full isotropic vector matrix. Fuller was an original, one of those do-it-yourself philosophers who don't flinch from reviewing the whole of human knowledge from their own angle and coming to their own conclusions. The result is admittedly idiosyncratic – when Fuller uses the word 'metaphysical,' for instance, he means 'weightless'; when he says something is 'rational,' he means it can be divided into 'whole-number increments.' Such an approach, ignoring even the philosophical traditions available to him, has its disadvantages. For one thing, his language remains opaque to most people. On the other hand, Fuller does manage to define his own terms in his own way, so that if you persevere, his way of speaking actually does come to make a good deal of sense.[20] It takes the 'synthetic' mind of a comprehensivist to see the principle of the whole(s) operating in every nook and cranny of the diverse natural sciences. Such minds are not likely to be fostered by the rigorous 'analytical' training that makes a scientist these days; they come to conventional science often as interlopers and mavericks, but also as illuminators.

One of the earliest and most important generalists who tried to see the world whole through the specialized lenses of twentieth-century science was of course Jan Christiaan Smuts, whose coinage of the term 'holism' in his now-classic *Holism and Evolution* (1926) was not restricted to the realm of biological evolution, although biology is the field where it has so far received the most attention. Smuts described his use of the word in the following way:

> *The close approach to each other of the concepts of matter, life and mind, and their partial overflow of each other's domain, raises the further question whether back of them there is not*

a fundamental principle of which they are the progressive outcome … Holism *(from Grk.* 'olos – *whole) is the term here coined for this fundamental factor operative towards the creation of wholes in the universe … The idea of wholes and wholeness should not therefore be confined to the biological domain; it covers both organic substances and the highest manifestations of the human spirit … There is an infinity of such wholes comprising all the grades of existence in the universe; … including all wholes which are the ultimate creative centres of reality in the world.*[21]

Arthur Koestler not so long ago took up Smuts' holism and refined it by positing what he called a 'holarchy,' that is, a hierarchy of wholes open-ended in both the *macro* and *micro* directions, each 'holon' coordinating those below it in the hierarchy while at the same time serving as an element in the larger whole above it.[22] If you add to Koestler's sketch the works of Bohm (the 'implicate order(s)'), Bateson (the 'pattern that connects'), Thomas Berry (the 'New Story'), Rupert Sheldrake ('morphogenic fields' and 'morphic resonance'), as well as René Dubos, Teilhard de Chardin, Michael Polanyi, Lewis Thomas and others, you begin to discern what can only be called a pattern or spectrum of patterns, connecting atoms to molecules to amino acids to proteins to cells to organs to organisms to minds to people to communities to cultures to bioregions to the living Earth to the galaxy at large … and, ultimately, some might say, to the communion of all beings in Being itself. We are once again approaching the *eukyklos sphæra,* the well-rounded sphere, of Plato's Pythagorean synthesis. Not surprisingly, if you look closely at Fuller's *Synergetics,* you may find that we have learned a good deal more about it over the intervening 2,500 years.

Much would become clear. You would recognize in your reflection that Descartes and Gauss had set up their XYZ coordinate system, still the basic graph for Western science, at deliberate variance from this 60-degree system. Through this grid, the entire natural world looks askew. You would see that even in the West, the 60-degree system was plainly evident – in cathedral architecture, for instance, in the coiled intricacies of Celtic knots – until the predominance of the square, cube and frame was firmly established by the rise of modern science. You would realize that the Renaissance 'perspective' of three-dimensional space projected into two dimensions by such a grid was a deliberate contrivance; convenient in some respects, but tending to isolate human consciousness (here, inside) on one side of the frame, segregated from the wide world (there, outside) on the other. Radical art historian José Argüelles writes:

The one-point, linear perspective is an intellectual, rationalistic, and above all purely mechanistic way of dividing space, a development synchronous with Gutenberg's press. Along with the printed word, it is the single most powerful agent for standardizing the perceptions of the collective mind of Europe over the next few centuries. If the printed word promotes mental uniformity, one-point perspective enforces that uniformity at the visual-sensory level. The grid system employed by one-point perspective is the forerunner of refined cartography and of the Cartesian system of coordinates, the source of all later scientific systems using graphs. Essentially, the one-point perspective is just that – a graph applied to the eye for the purpose of mechanizing vision, and thus mind.[23]

You would not only to begin to inveigh, as Bucky did, against squares, cubes, and the blockheads who thoughtlessly employ them – as arbitrary, abstract, misleading and inefficient models of Nature's own very definite, concrete, and efficient structural strategies – but you would perhaps begin to notice that in cultures where this Western 'viewpoint' had no foothold, Nature's co-ordinate system still holds sway.[24] Anywhere Euclid, Descartes and the abstract analytic geometry they spawned never took hold, you tend to find patterns allied to those Fuller himself developed – with both the technical resources and the pragmatic biases of a twentieth-century scientist-engineer-artist.

As a single example, consider one of the earliest instances of 60° radial symmetry: a striking Harrapan bas relief found in the Indus Valley, dating back at least 4,000 years.[25]

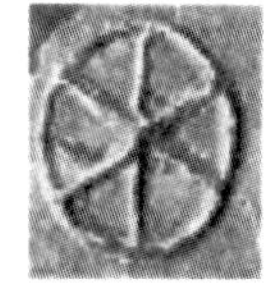

Karachi Mus.

Its six-way division of the circle exactly mirrors the great circles slicing through the Vector Equilibrium; or else, given its antiquity, Fuller's figure is in fact little more than an extremely precise version of that venerable carving. Besides later serving in iconography throughout India as the graphic figure for the second (genital, generative) *chakra* – described in the literature as 'energy's own standing space' – and as the model for innumerable common *yantras* (meditation diagrams), its circular 60° symmetry may well have provided a prototype for the hexagonal architectural symmetry of many later Hindu temples. It would take us fairly far afield to detail further the myriad correspondences of Fuller's geometry with the Minoan *labrys,* or the Pythagorean *tetraktys,* or the *Kabbalah's* Tree of Life, not to mention the patterns in Persian carpets, or Celtic knots, or cathedral rose windows, but all of these pre-Cartesian crystallizations of the isotropic vector matrix and its characteristic 60° symmetry are undeniably tangled up with the very roots of Indo-European culture.[26] And, not altogether surprisingly, with those of very many other cultures as well ... Indeed, once you start looking for Nature's coordinate system in the iconographies and sacred geometries of traditional cultures, you can scarcely avoid discovering antecedents everywhere. A scattering of illustrations will suffice to underscore the point.

As a further boon, you might begin to look at Bucky's domes in a very different way, as vast figures for contemplation, aerial traceries etched against the vault of the sky. Many of Fuller's constructions are not primarily utilitarian, but thought-provoking, their lively interplay of positive and negative spaces evoking 'skyviewing rapture,' as Bucky's lifelong friend, the Zen-Modernist sculptor Isamu Noguchi might express it. The American Society for Metals Dome in Cleveland, for instance, has no ostensible purpose beyond showing off what can be done with metals. The playdomes in the childrens' playground are not for work but for play. The Climatron in St. Louis encloses a miniature biosphere that simply exists, and that is its statement: life nurtured under Bucky's umbrella. As a matter of fact, this germinal idea has recently sprouted up elsewhere. The Eden Project in Cornwall, Britain's 'living theatre of plants and people' (which opened to visitors early in 2001), consists of a series of giant interlocking geodesic conservatories called 'Biomes,' both extending and beautifully refining Fuller's original Climatron concept.[27] And of course there is little left of the Montréal Expo Dome but its enigmatic skeleton, yet reflecting on that alone has taken us places we might not have expected.

I would like to think that if we are sensitive to this contemplative dimension of Fuller's constructs, we may begin to recover one of the eldest domains of sacred geometry, *the mandala:* the world as pattern, and that pattern as an intelligible world.

Fuller with early geodesic dome, dymaxion map, tensegrity structure and models derived from his energetic/ synergetic geometry; Forest Hills, New York, 1951. (Estate of R. Buckminster Fuller)

American Society for Metals Dome, Cleveland. (© Robert Duchesnay, 1990)

Climatron, St Louis. (© Robert Duchesnay, 1990)

Climatron, St Louis. (© Robert Duchesnay, 1990)

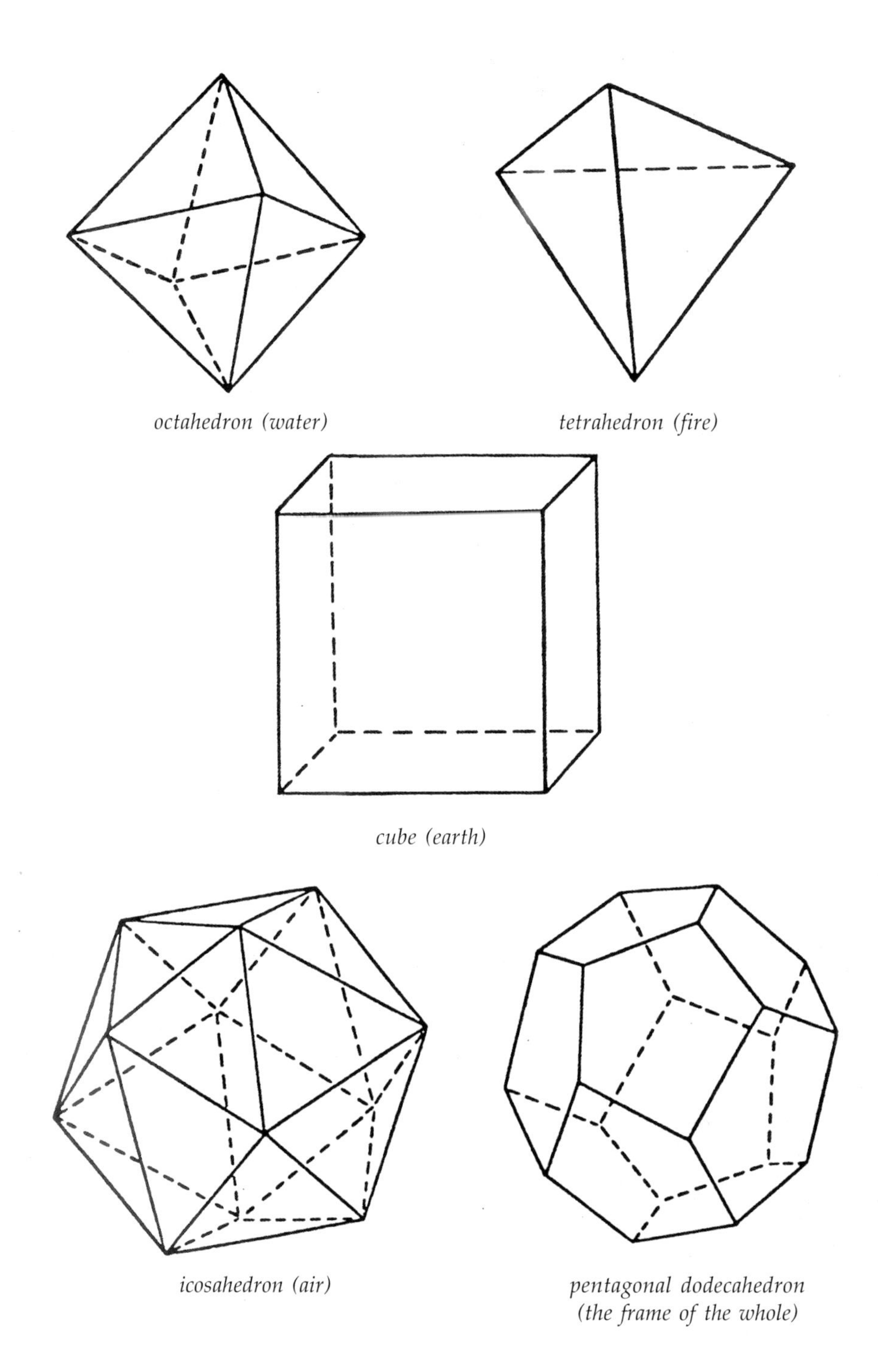

The Five Platonic 'Solids' (as 'elements' in the Timaeus)
(Ghyka, Geometry, *NY, 1946)*

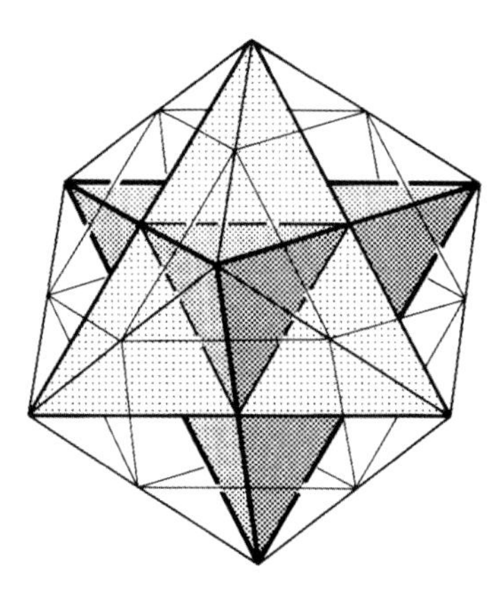

Fuller

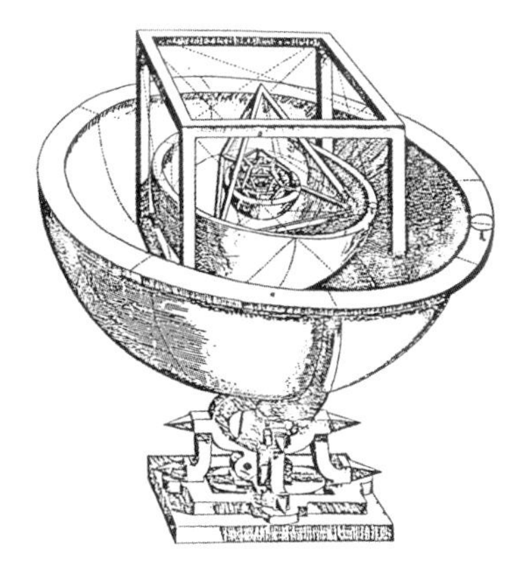

Kepler

Ardagh Chalice (detail)

Śri Yantra

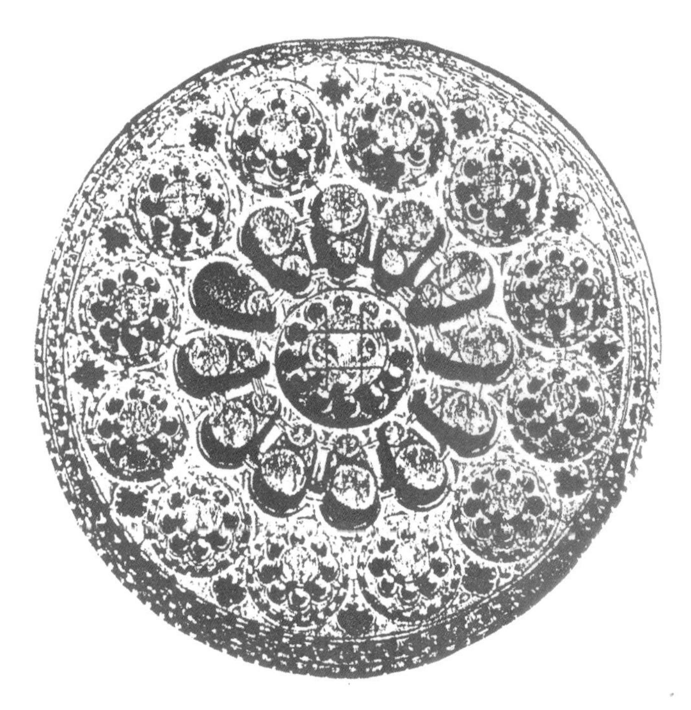

Rose window from Chartres Cathedral

Labyrinth from Chartres Cathedral

Arabic Geometrical Motifs

Mandala

Deep South

Adding his usual descriptive flourishes, biographer Alden Hatch relates one of Buckminster Fuller's adventures which has in retrospect very nearly become legend:

> *In 1958, on his first trip around the world, Bucky stopped off in India where he had numerous speaking engagements. One day he made three speeches in New Delhi, in the morning, afternoon, and evening. At all three meetings he noticed a striking youngish woman dressed in exquisite saris, who usually sat in or near the front row listening intently, her large dark eyes refulgent with intellectual excitement.*
>
> *At the evening lecture, Bucky used a little tensegrity sphere model, only about six inches in diameter, made of string and small turnbuckles. After the lecture he was presented to the lady, Mrs. Indira Gandhi, Prime Minister Nehru's daughter. She was so interested in the model that he gave it to her. Mrs. Gandhi asked him if he would come on Saturday to meet her father at their house. Bucky says, "I did meet him and we had a very extraordinary time" …*
>
> *Just what he said that day no one will ever know, for it was, as always, purely spontaneous … In the time he allotted himself he endeavored to give Nehru the essence of his philosophy and the logical reasoning on which it was based. During the entire period Nehru stood quietly intent, absorbing Bucky's words.*
>
> *At the end of an hour and a half Bucky stopped talking … Nehru bowed with folded hands and left the room. He had not spoken a single word.*[28]

An extraordinary time, indeed … *not a single word?!* Bucky had remarkable powers as a monologist, granted, but by the same token a signal disinclination to dialogue. (Part of the problem was his deafness; he could not hear others well, and the phrases he did hear tended simply to trigger off a new monologue; but that's another issue). Bucky was a man possessed by his own vision. Of course he had several, more conversational meetings with Nehru, who had read and admired his books, and with Indira Gandhi too, who eventually invited him to design the Delhi airport, and to give the Nehru Memorial Lecture in 1969 during her own debut as Prime Minister. They remained lifelong friends, Bucky ever unwilling to concede that the realities of weilding power might have changed her over the years from the eager, "intent" young woman he had met in 1958. But I would like to take this first meeting as a symbol for something beyond personalities, beyond politics for that matter: the encounter of East and West strangely symbolized by Nehru's silence before Bucky's customary avalanche of words – the silent bow of the 'under-developed' India before a distinctly Western vision of her prospects for technological development.

For surely that is what Bucky outlined, and what Nehru was all too ready and willing to receive. His 'design science revolution' was the constant theme of all Fuller's talks and books, but here might have received an Indian twist, as in his "Ten Proposals for Improving the World":

> *The intellect, vision and courage of Mahatma Gandhi conceived of passive resistance with which bloodless revolution he broke the hold on India of history's most powerful sovereignty.*

[But ...] passive resistance will not amplify the production of life support.

In extension of the Mahatma's magnificent vision we are committed to the design science revolution by which it is possible bloodlessly to raise the standard of living of all humanity to a higher level of physical and metaphysical satisfaction than hitherto experienced or dreamed of by any humans. All the knowledge and resources are now available with which to accomplish that comprehensive success.

This can all be realized by 1985 [!] without any individual profiting at the expense of others or interfering with another's enjoyment of total planetary citizenship. It requires the competent design science commitment of world-around youth to realize the Gandhian integrity.

Youth's spontaneously mutual commitment can only be inspired through experientially gained knowledge and love-sustained innate faith in the eternal reliability of cosmically manifest principles as discovered by science.[29]

This "extension of the Mahatma's ... vision" is vintage Bucky, of course, not Gandhi at all ... except for the "bloodless" non-violent ideal they shared (though Nehru and Indira Gandhi apparently did not). One notes right away that Bucky imagines "the Gandhian integrity" will be realized through "principles ... discovered by science," not by a return to the Indic traditions which nurtured Gandhi, nor to Gandhi's own proposals for agrarian reform, rural self-sufficiency, hand-spinning and weaving and other village crafts. Gandhi claimed he was not against technology, yet his idea of 'appropriate' technology was entirely pre-industrial: "The spinning wheel itself is a machine. What I object to is the craze for machinery, not machinery as such ... The supreme consideration is man." In the same epoch as Fuller, you may recall, E. F. Schumacher proposed to revive the Buddha's suggestion that everybody in India plant one tree a year for five years, a notion which seems much more along the lines of Gandhi's own approach than the airport Bucky designed for New Delhi. (Today that proposal would produce something like five billion trees – a new world!) Gandhi's way, in tune with the rooted traditions of Indian peasant life, was in fact profoundly out of tune with the programs of massive industrialization his successors undertook in his name, leading eventually – might one even say inevitably? – to India's own atomic bomb, and then to Pakistan's, and so on.

The doctrine of development had been outlined by US President Truman only a decade before Fuller's visit to India, in his Inaugural Address of 1949. At a stroke the world was divided into rich and poor nations, North and South, developed and 'under-developed' areas – based solely on GNP. And the doctrine was readily accepted on both sides of the 'development' divide, across which monies flowed freely for a time. Had Nehru spoken that day, he might well have lamented India's artificially 'arrested' technological development, as he had elsewhere, and laid the blame directly at the feet of the British. It was no coincidence, he would assert, that the poorest Indian provinces were those that had been under British rule the longest. In his *Discovery of India,* written from Ahmadnagar Fort prison in the early 1940s, Nehru drew attention to the coincidence of the English industrial revolution in the late eighteenth century with the systematic looting of once-wealthy Bengal, which provided vast reservoirs of ready capital for the new industrialists. Their policies were to change the entire relationship of Britain to its Indian colony:

The chief business of the East India Company in its early period, the very object for which it was started, was to carry Indian manufactured goods – textiles, etc., as well as spices and the like – from the East to Europe, where there was a great demand for these articles. With the developments in industrial techniques in England a new class of industrial capitalists rose there demanding a change in this policy. The British market was to be closed to Indian products and the Indian market opened to British manufactures ... This was followed by vigorous attempts to restrict and crush Indian manufactures ... The Indian textile industry collapsed ... The process ... continued throughout the nineteenth century, breaking up other old industries also, shipbuilding, metalwork, glass, paper, and many crafts ... Machinery could not be imported into India. A vacuum was created in India which could only be filled by British goods, and which also led to rapidly increasing unemployment and poverty. The classic type of modern colonial economy was built up, India becoming an agricultural colony of industrial England, supplying raw materials and providing markets for England's industrial goods.[30]

What Nehru probably saw in Fuller was a fast-track to the most futuristic technology imaginable. What Fuller offered was in fact a maverick technology entirely at variance with the monotechnic trends of Western industry and capital development, envisioned as we have seen as a series of interlocking service industries for all humanity, rather than more profit-taking enterprises for the benefit of a few Great Pirates. Bucky was advocating something very much like what is today called 'sustainable' development or 'green' redevelopment, but Nehru's openness to modernization brought mainly the mega-projects and social engineering typical of development programs the world over, and typically unsuited to the indigenous people or their ways of life. While it is surely true that Nehru did not need Bucky to teach him about Great Pirates and their colonial policies, the disaster of Western-style technological development in India over the past fifty years testifies that development itself has turned out to be little more than the reincarnation of the very colonialism Nehru struggled against. His celebrated foreign policy of 'non-alignment' with either the Soviet bloc or the Western bloc did not prevent him from aligning his people with a modernization which inexorably tried to Westernize India. So today, the rise of Hindu fundamentalism, the increasing sectarian violence, and the continued 'modernized' poverty, are the bitter legacies of such development. Writing a decade ago from New Delhi's Center for the Study of Developing Societies, Ashis Nandy comments on the demise of the twentieth century's dream of science and technology as a liberating force beyond politics:

The earlier creativity of modern science, which came from the role of science as a mode of dissent and a means of demystification, was actually a negative force. It paradoxically depended upon the philosophical pull and political power of tradition. Once this power collapsed due to the onslaught of modern science itself, modern science was bound to become, first, a rebel without a cause and then, gradually, a new orthodoxy. No authority can be more dangerous than the one which was once a rebel and does not know that it is no longer so.

The moral that emerges is that modern science can no longer be an ally against authoritarianism. Today it has an in-built tendency to be an ally of authoritarianism. ... the scientist has become the main author of the establishment cosmology. He is now the orthodoxy; he is now the Establishment. So much so that to perceive him still as a weak, unorganized fighter against authority can spell disaster for all of us ...

> *It is therefore not a paradox of our times that to contain modern science many are falling back on what has been one of the main targets of modern science during the last three hundred years – cultural traditions ... This, however, only brings us to another question: What kinds of tradition can be used as tools of criticism and what kinds are open to criticism?*[31]

In any case, neither Fuller nor Nehru (a brahmin who confessed himself a poor student of the Indian religious classics) paid much attention to the wisdom traditions of India, let alone to her architectural and iconographic heritage. Nehru and his daughter looked to the West, and to modern history with its arenas of politics, law, science and technology, military and police powers. Setting up their own sovereign democratic nation on the Western model, the new rulers of India may well have sought, in Nandy's phrase, "to pay back the imperial West in its own coin." Fuller had a good grasp of history, and the history of the British Empire in particular (he followed the exploits of seafaring nations with keen interest), but virtually no background at all in the traditions of India. Beyond this, and probably because he was too busy living out the consequences, he seemed unaware of the power of modern science and monotechnics to co-opt and absorb mavericks like himself into orthodox avenues. As Wolfgang Sachs notes, Truman's Inaugural Address "called for technical assistance designed to 'relieve the suffering of these peoples' through 'industrial activities' and 'a higher standard of living.'"[32] Overseas, Fuller must simply have sounded like a charmingly idiosyncratic Yankee proponent of the cause; a happy propagandist who thought it was all his own idea.

Of course with Nehru, Fuller was preaching to the converted. Nehru believed strongly in the underlying unity of India, in her ability not only to tolerate but to absorb and domesticate foreign influences. Of India's past, he once wrote: "Foreign influences poured in and often influenced that culture and were absorbed. Disruptive tendencies gave rise immediately to an attempt to find a synthesis. That unity was not conceived as something imposed from outside, a standardization of externals or even of beliefs. It was something deeper ..."[33] Yet with his uncritical acceptance of modern industrial development, Nehru was letting in not Fuller's relatively benign 'design science revolution,' but what turned out to be a mega-technic Trojan Horse.

At any rate, nobody in the picture at the time seemed to have either the foresight or the hindsight to look at Fuller's geometry through Indian eyes. It is an astonishing lapse, since the correlations with Hindu iconographic forms are so obvious. I can however tell you from personal experience that Buckminster Fuller, friend and confidant of two Prime Ministers of India, had never examined a *śri yantra* until I had the opportunity to show him one in mid-1983: "That's an isotropic vector matrix," he exclaimed straightaway, a note of surprise in his voice. And so it is, although slightly skewed – as if viewed from a certain angle and projected onto a flat plane. One might even speculate that this is a view of the isotropic vector matrix *from the inside.* But what does it mean that the core matrix of Fuller's geometry should show up so accurately rendered not only in this most revered and prominent *yantra,* but in almost every other Hindu meditation diagram, and in the ground-plans of so many Indian temples as well?

Thereby hangs a remarkable tale which Bucky simply did not live long enough to tell. Will you permit me to make up some of this untold story as we go along?

Due East

First of all, what exactly are mandalas? *Mandala* is simply the sanskrit word for a circle and center. It was traditionally reserved for Buddhist meditation diagrams, while their Hindu counterparts were called *yantras,* but today the word *mandala* has become a generic term covering all such sacred spaces from the rose windows in cathedrals to Navajo sandpaintings to the enigmatic pictures produced by psychologist Carl Jung's patients.[34] The short answer is that mandalas are *symbols of the whole.*

In modern, secular society, it may be that more questions are raised than answered by such a definition. What, for example, is a symbol? I shall go out on a limb here: a symbol is the real appearance of something real, which would not otherwise appear at all. Drawing on R. Panikkar's analysis, I should add that a symbol is neither a purely objective entity (in the world, out there), nor a merely subjective entity (in my mind, in here), but precisely the bridge between object and subject, outer and inner, the real and the ideal.[35]

And in this day and age the very notion of *the whole* is problematic too, in ways our forebears might never have suspected. We may well have our own idea of the whole – it is all water, or fire, or matter, or God, or consciousness, or chaotic or absurd or whatever – or we may concoct a marvelous system and claim it embraces everything, but we are all too well aware there are others with other ideas, and other systems which also claim to be both comprehensive and comprehensible. We can no longer sustain the pretense to universality of any one vision of the whole – which we know must be conditioned by a specific nature, culture, language, religious tradition, and whatever other factors we deem important. Are we then to suppose that there is no whole, no underlying unity at all? That we are all locked up in our private worlds and images and language games, and that there is no escape? But such an 'absolute' relativism contradicts itself ('Everything is relative except this statement') and, strictly speaking, cannot even be communicated. If there is no larger whole (again, of nature, culture, language or what have you) in which we both to some degree participate, then I am merely talking to myself here and you are hearing only some distant echoes of your own voice in your head.

It is precisely in bridging this apory between the One and the Many, as well as between Self and Other, that symbols in general, and mandalas in particular, come to our aid. A mandala is not primarily a picture of anything. Mandalas are specifically designed to integrate and resolve apparent oppositions and conflicts into the complementary facets of a more comprehensive understanding. The paradoxes of human life do not disappear, but they are transformed – shall we say transfigured? – in the light of the whole. The process of making a mandala brings one face-to-face with the patterned integrities of life itself: center, symmetry, and rhythmic resonance. And there is more: "The larger whole," as Gregory Bateson once phrased it, "is primarily beautiful."[36]

Anyone with half a wit can see patterns in the flowers, can sense the rhythm of the ocean waves, can marvel at the orderly tracks of the planets through the starry heavens. True, we cannot 'grasp' the whole of reality, we cannot 'know' it as we know our phone number, nor claim it as our own idea. Of course not. There is an obvious mystery here, as we earlier found both Fuller and Mumford affirming, but also a plain and simple truth. Although we rarely notice it, we can and we do understand something very basic about the entire reality of our experience: *it coheres.* It seems to 'work' – it 'fits' together quite splendidly; it is whole, intact, sound. It displays, as Bucky tirelessly reiterated, its own *integrity.* Yet that integrity is not just some objective datum lying about somewhere. It is integrally tied to our own integrity, our own sense of what is true and real. And it is precisely this link between our awareness and the world of which we are aware that is celebrated – indeed, meaningfully 'cerebrated' – by the mandala.

Certainly no single cultural product, no single format or medium, could claim a monopoly on the global mandala tradition. In bygone days, to found a city you laid it out in a mandalic pattern that mirrored the very structure of the universe, or else it wasn't even a 'place' at all.[37] To build anything – a home, a city, a pyramid or cathedral, a stone circle or burial mound – you closely followed the iconographic tradition you had inherited because otherwise what you built would not even be *real,* let alone endure the passage of time. Yet as they have come down to us, the classical mandalas and yantras and masonic diagrams have mainly been confined to two-dimensional representations.[38] However much depth they are intended to suggest, they remain rather flat and static. Bucky Fuller's constructions, taken as figures for contemplation, would appear to be a quite natural evolution of this symbolic heritage. They simply 'round out' the traditional mandalic structures and allow us to feel our way into their labyrinthine depths.

Such associations may do two things. They may provide Fuller's geometry a traditional context beyond the purely scientific: in a word, roots.[39] And Bucky's reflections upon such figures may, in turn, allow the sacred geometries of many a traditional culture not only to retain their relevance to lived human experience, but perhaps even to shed some light in a realm previously off-limits to them: modern science itself.[40]

The impact of modern science on most traditional world-views has been quite simply and drastically to dissolve them, to dissipate their vital centers of value; its effect has been *entropic.* Our order disorders theirs, so to speak, and tends to leave traditional cosmologies in disarray. But the search for meaning goes on. Even a theoretical physicist of the stature of Roger Penrose doubts that the equations for astrophysics and those for quantum mechanics will ever be meaningfully connected. Yet between macrocosm and microcosm lies the mezzocosm of the human condition, where most of us live out most of our days. One way of looking at Fuller's synergetics might be to see these as patterns constructed out of all this detritus floating around, occasioned by the entropic dissolution of traditional patterns of meaning and value. The disintegration of so many traditions would, according to Bucky's understanding of synergy and entropy, have to be taken up somewhere else, into some other system of meaning and value. And science itself is in many respects the dominant 'religion' of our time, a belief system – the word is *scientism* – so ingrained for most people that it is simply taken to be real and not a matter of belief at all. Every astrophysicist worth his or her salt today has concocted their own cosmology. But these are mathemathical projections, not lived worldviews, not meaningful universes within which human beings have spent millennia working out the patterns and symbols and ways of life which correspond to their structures and render life within them worth living. Even the few symbolic systems we have riffled through here are enough to suggest that Buckminster Fuller's cosmology may turn out to be the exception.

The emergence of Fuller's energetic-synergetic geometry – its remarkable predictive power in so many scientific fields coupled with its striking resemblance to a myriad of traditional cosmological systems – may yet turn out to be a seedbed for the eventual transformation of the modern scientific/technocratic worldview into something more life-enhancing, if not indeed more real. Here is Fuller on this Janus-faced topic:

> *Order and Disorder: Birth and Growth: Entropy is locally increasing disorder; syntropy is locally increasing order. Order is obviously the complement, but not mirror-image, of disorder ... Universe is a vast variety of frequency rates of eternally regenerative, explosive, entropic vs implosive, syntropic pulsation systems. Electromagnetic radiant energy is entropic; gravitational energy is syntropic.*
>
> *Both entropy and syntropy are operative in respect to planet Earth's biospheric evolution. Wherever entropy is gaining over syntropy, death prevails; wherever syntropy is gaining over entropy, life prevails.*
>
> *Entropy is decadent, putrid, repulsive, disassociative, explosive, dispersive, maximally disordering, and ultimately expansive. Syntropy is impulsive, associative, implosive, collective, maximally ordering, and ultimately compactive. Entropy and syntropy intertransform pulsively ... There is an entropic, self-negating, momentary self; there is also the no-time, nondimensional eternity of mind ... Every time we experience the ... disconnects of momentary annihilation into eternity, naught is lost. Mind deals only with eternity – with eternal principles. What is gained to offset any loss is the residual, observational lags in accuracy inherent and operative as cognition and the relativity of awareness that we call life.*
>
> *The life-propagating syntropy-entropy, birth-to-death transformations constitute the special case realizations of the complex interactive potentials of all the eternal, abstract, dimensionless, nonsubstantial, generalized principles of Universe, interplayed with the 'if-this-then-that' integrity of plural cosmic unity's intercomplementarity. The death and annihilation discontinuities occur as eternal generalization intervenes between the special case, 'in-time' relative intersizing of the realizations.*[41]

Such an understanding, if you can penetrate Bucky's idiolect, accords astonishingly well with traditional understandings of the mandala as both a figure for the cosmos at large and an intimate map of the soul's journey to understanding it.

Moving outward from the center, the mandala begins with the simplest elements and moves through level after level of life and being until it encompasses every aspect of daily life. At the center may be a point *(bindu)*, or a God, a geometrical figure like the yin/yang polarity, or maybe a sexual union. Usually the archetypal Word in one form or another is manifest *before* the myriad things of this world: strophes and melodies, the sacred hymns, prayers and invocations take precedence in this chain of being, because it is first and foremost a chain of meanings. At the next stage, in a primal explosion of light and life, you might see angels or boddhisattvas balancing on this lynchpin, then the sun, moon, stars and planets, then come all the plants and animals, and finally perhaps, the human world and some everyday scenes of domestic life. The entire order of created being radiates outward from the one central source to the circumference.

Contrariwise, moving inward toward the center corresponds to the experience that there are levels of consciousness, and degrees of intelligibility: everyday sense awareness first; then strong emotional colors, perhaps; then rational divisions into right and left, up and down, light and shadow, etc.; next the realm of imagination where the highest Gods are at play in the human psyche; then, even higher and more simple, the intuition of the basic principles ("eternal, abstract ... generalized principles"?) by which the whole holds together;[42] and finally a merging with the very core of life in which the individual 'self' is annihilated by the discovery of its true 'Self,' perhaps, and so on. From the One to the Many, and back again; from birth, through growth and fullness, dissolution and death, then back to life again.

Fuller's Universe, too, was an eternally self-regenerate 'scenario,' not the one-shot, open-and-shut book so popular in today's science, with its 'arrow of time' running from a dateable beginning ('Bang') to a predictable end ('Crunch'), though Fuller might accept this as one 'pulse' in the larger rhythm. Here the creativity of the creation is a constant, which may be tapped at any and every moment; the source of Life may be glimpsed in its endless resourcefulness, in the primordial interplay of syntropy and entropy, fullness and emptiness, systole and diastole, the rhythmic breathing – in and out, in and out – of the entire reality. And the proper way to 'view' a mandala? In the Navajo sandpainting rituals, the subject sits in the center.[43]

Outward from the center the mandala is cosmography, or indeed, cosmogenesis: the centrifugal, explosive unfolding of life in all its expansive 'outer' aspects. Inward from the periphery the mandala forms a 'psychograph,' the soul's symbol of its own inner growth: the awareness that is the centripedal, implosive 'inner' dimension of depth. And at the center? That's the mystery some traditions would call divine, but it need not be couched in theistic language. Need we indeed say anything at all about this rhythmic interval, the empty center (that is clear), the nowhere (that is everywhere) between the *ex*-plosion of life and the *im*-plosion of awareness? Maybe that dimension is best characterized as nothing but a simple 'plosion' – from *plaudere,* 'to clap hands' – or else the sound of one hand, clapping.

At any event, the mandala bridges the abyss between inner and outer, subject and object, humans and the natural world, that has so long been sundered in our Western, Cartesian world-'view.' In the Neoplatonic procession (or emanation) and return of all things to their Source – the very dynamism of trinitarian theology expressed from the Pseudo-Dionysius and Scotus Eriugena through Aquinas and Bonaventure to Meister Eckhardt and Nicolas of Cusa – we see the same pattern front and center in Western thought before it was cut short by the scientific reductionisms. Bucky Fuller's geometry is a *novum* in the scientific world precisely because it moves toward restoring this same coincidence of opposites – of inner and outer, cosmos and consciousness – and attempts to see in them and through them a common integrity: "Unity is plural," says Bucky more than once, "and at minimum two."[44] Fuller seems to have anticipated the recent 'eco-psychology' movement (of Theodore Roszak *et al.*)[45] by at least two decades. If nothing else, the mandala serves to remind us that *relationships* are at least as real as whatever they are said to relate.

Bucky insisted that model-making was the only way to really enter his geometry of mind and nature, and he was right. Before the abstract digital world of the computer entirely erodes our manual sense of contact with real *things,* we might consider taking his advice.[46] He originally wanted to begin *Synergetics* with what is now Chapter 8, "Operational Mathematics," presumably so that people would have some hands-on experience with the models before struggling with his prose. Model-making does actually make the writing a good deal clearer; you get a feel for the things Fuller is talking about. *Synergetics* as it stands still contains the specifications for some models. Extensive commentaries from Anthony Pugh, for instance (who has done much work on the tensegrities Fuller developed with sculptor Kenneth Snelson back in the Black Mountain College days),[47] or Hugh Kenner (on geodesic math), offer plenty of others. Fuller's models – and for me, I might add, particularly the foldable great-circle models in *Synergetics* (the 'WHOLES' of Appendix A) which highlight the dynamic relationship of center to circumference that we have been discussing – provide much the same direct experience with matter itself that more traditional materials have always afforded the artisan. You discover matter in its particularity, the diverse properties of various materials, the constraints of time and space, and most importantly, which principles 'fit' what practices (and which do not). The construction of a mandala was – and for that matter still is, in many cultures – a ritual act, which is to say an act of 'world maintenance,' upholding the *ṛta,* the 'rite' or ritual order (*ordo,* the 'weave') of things as they are. Bucky, of course, was never just building a dome; he was trying to save the world. The right attitude is a contemplative one, maintaining at once an ultimate seriousness toward whatever one is doing, and a playful sense of abandoning oneself to it. Here we take our cues from the artist. As Eric Wesselow reminds us, "There are no minor arts, only minor artists."[48] In the same vein, he passes on the parable of the stone mason, itself passed down for God knows how many years:

Once upon a time
there was a king who wanted
to know more about his people.
So he dressed as a wandering merchant
and visited many towns and villages.

One day he stopped at a site
where a building was going up.
He asked one of the workmen,
'What are you doing?'
The man answered,
'I am cutting stone.'

Then the king asked another man
who was also cutting stone,
'What are you doing?'
And the second man answered,
'I am building a cathedral.'[49]

True North

Patterned integrities, along with the human ability to discover, articulate and understand them, have long fascinated our species. The oldest known human artifact, an eight-spoked 'sun-wheel' found by Louis Leakey in Olduvai Gorge, attests the longevity of this concern. Civilizations may rise and fall, but their patterns remain ...

Spherical Tetrahedron (by S. Eastham; photo © R. Duchesnay, 1990.)

From the *T'ai Ch'i* (yin/yang) diagrams of ancient China, to the rose windows of medieval Europe, to the *vajras* and *dorjes* (thunderbolts) of India and Tibet, to the magic carpets of the Middle East, these sacred geometries flourish with astonishing vitality and diversity. Many are still living traditions, able to adapt and transform themselves in the rapidly changing circumstances of the modern world. We come to see that such figures are not isolated 'art objects,' any more than Fuller's synergetic geometry is some remote mathematical abstraction. It is all an incredibly precise metaphor, delimiting what could be called a symbolic spectrum, a keyboard of the imagination, a range of formal possibilities on which the variations and permutations may well be endless.

It is important to note the limit cases. As Amy Edmonson nicely puts it in describing Fuller's work, "Space has shape."[50] It is not a Newtonian blank, but an articulated emptiness. Fuller himself speaks of the "12 degrees of freedom," which he derives from 'opening out' the six radii of the vector equilibrium – seen as both expanding outward from the center, and contracting inward.[51] In drawing up the specifications for the foldable geometries in *Synergetics*, he notes that only seven such great-circle figures

(WHOLES) are possible, since they are derived from the 'seven axes of symmetry' offered by spinning the tetrahedron, octahedron and icosahedron on their respective vertices, mid-edges, and face-centers.[52] These are cosmic limit-cases, like the 92 regenerative chemical elements, which Fuller maintains are tied to the vector equilibrium configuration achieved by the closest packing of spheres around a central sphere. The tetrahedron itself is a limit-case; nothing simpler has an inside and an outside. The tetrahedron ("the minimum structural system in Universe," says Fuller) gives you maximum surface and minimum volume, while the sphere ("a plurality of energy events approximately equidistant from a central event," says Fuller after contructing thousands of them) is at the other end of the scale, maximum volume for minimum surface. My own symbol for this entire range of symbolic possibilities is an open spherical tetrahedron, a kind of sculptural oxymoron. It underscores Fuller's seemingly odd claim in *Synergetics* that the tetrahedron is the "minimum sphere."[53]

Les extrêmes se touchent! On the popular 'Science Solves Ancient Riddles' front, Fuller's geometry apparently resonates equally well with scientific futurists and archaeological explorers of the past. Looking forward, Pat Flanagan of 'pyramid power' fame wrote another book, *Beyond Pyramid Power*, in which he cited Fuller's isotropic vector matrix as the shape of "bioplasmic energy fields" about to be uncovered at the frontiers of modern science.[54] Fairly far out, of course, but Flanagan did do some empirical testing difficult to explain away. Looking backward, Peter Tompkins proposed in his *Mysteries of the Mexican Pyramids* that Fuller's geometry holds the key to the mathematics employed by the builders of the sun and moon pyramids of Teotihuacan – extrapolating a bit freely, perhaps, but based on the work of legitimate field researchers:

> *Harleston concludes that the fundamental message conveyed by the Teotihuacanos is that the physical universe is tetrahedral from the microscopic level of the atom all the way up to the macrocosmic level of the galaxies, on a scale of vibrations in which man stands about the center. Man would thus have built into him, as suggested by Pythagoras and Plato, the tool for unlocking the geometry of the cosmos and recovering the knowledge of his role in the scheme.*[55]

Fuller's own elaborations of the synergetic geometry include several such hierarchies, closely paralleling the inward/outward thrust of mandala meditations.[56] The most striking single figure may be his "Cosmic Hierarchy of Omniinterrationally-phased, Nuclear-centered, Convergently-divergently Intertransformable Systems," one of the color posters Ed Applewhite has included as an extended frontispiece to *Synergetics 2.*

A more elaborated sequence, less abstract and more accessible to the non-geometer, is to be found in the "Cosmic Hierarchy of Comprehensively Embracing Generalizations" in the first *Synergetics* volume (¶1056.10). Here Fuller begins with "the cosmic intellectual Integrity manifest by Universe"; moves down through Synergy (symbolized by a *T'ai Ch'i* diagram!); Nature, Unknown and Known, metaphysical and physical; syntropy and entropy; astrophysics; the solar system, Earth and its "biologicals"; humanity with its various philosophical and political subdivisions; and finally, Them, We, You, and a full dozen and a half dimensions of "Me." Despite the souped-up scientific language, the reflection moving from the outer cosmos to inner consciousness has much the same feel as many a traditional meditation on the inter-related dimensions of the mandala. A more traditional language would say that Fuller has found his way to the central axis of the world soul. Like the shaman clambering up the cosmic tree, or some latter-day Dante returning to this world after an ascent into the celestial spheres, Fuller travels freely from dimension to dimension via this pivotal *axis mundi*.[57]

Permit a final pictorial 'word' about those mandalas. In the following two illustrations, compare for yourself the triangulation of the Hindu *Śri Yantra* with the pattern used to derive the canonical proportions of the Buddha-body. It is most curious that the Hindu tradition – which fills in the doctrine of the essence (*atman*) of all things with abundant detail – should choose to empty its most prominent meditation diagram; while the Buddhist tradition – which asserts the emptiness (*śunyata*) of all things – has opted to fill in the same or a remarkably similar triangulated matrix with the body of the Buddha.

Empty or full, much the same coordinate system seems to have captured the imaginations of both Tantrist and Buddhist monks, gainsaying the Himalayan peaks of orthodoxy which so often seem to stand between them.

Here again, then, are the skeletal triangles of the classical *Śri Yantra*, as depicted in Tucci's *Theory and Practice of the Mandala:*[58]

In deliberate contrast to the rich fullness of the *triloka*, the 'three worlds' of the Hindu tradition, the Buddha in his radical mutation of Indic religiousness advocated the 'Eightfold Path' and an awareness of the 'emptiness' of all forms. Could the 'diamond body' of Buddhism not have something to do with the octahedral voids which inevitably emerge between tetrahedra in an isotropic vector matrix?[59] Are such 'structural' reflections permissible?

Here are the canonical proportions of the Buddha-body, as depicted in Singh's *Himalyan Art*:[60]

You are free to draw your own conclusions.

Way Out West

It is perhaps enough to have glimpsed the structural convergences, without trying to minimize the cultural divergences, between these traditional patterns and Buckminster Fuller's "geometry of thinking." But before we too glibly pronounce the union of tradition and modernity a happy marriage, we must recognize that the rift we discerned earlier – between Lewis Mumford's determination to cling to cultural roots and Fuller's eagerness to go out on unexplored limbs – has not only deepened for subsequent generations, but has also, in some respects, changed form. Despite millions of homeless and displaced people the world over – more today than ever before in human history – housing and architectural style is not at the forefront of the world's agenda.[61] Despite the prevalence of technology, and its disruption of traditional ways of life, the entire direction of modern technology is simply accepted and the impact of that technology on human values (or vice-versa!) has been left in the shadows – as when a multi-million dollar university complex for genetic engineering is built in Maryland, and a few dollars promised (one day) for an institute to consider its ethical ramifications.[62] No, today actual human problems seem to have taken a back seat to 'virtual' realities in the new computerized world of the Internet and the World Wide Web. We seem indeed more concerned these days about the map than the territory, more exercised about the latest computer simulations than about the structure of reality.

And yet the dichotomy which polarized Fuller and Mumford has never been resolved, and leaves us more ambivalent about it all than ever. The new media technologies seem to be bifurcating people into two mutually irreconcilable time zones: the *futurists*, those who are – sometimes cautiously, sometimes eagerly – looking forward, aching to live in this hi-tech, hi-touch future-world; and what I would call the *archaists*, in the best and original sense, those who are – sometimes nostalgically, sometimes polemically – looking backward toward the origins, the *archai*, preferring to hold onto whatever is left of the past, for all its faults and foibles.[63]

The prevailing problem is an extension of the dilemma that faced Fuller and Mumford, precisely the inherent universalizing tendency of global media technologies, well beyond the bounds of the Eurocentric culture which originally conceived them. We are fast tending toward one global communications system. In skeletal form, it exists already. Yet no matter how 'pluralistic' such a system may seem to its designers, it is increasingly evident that many other cultures do not and will never find it so. They find instead their rich and unique cultural universes swiftly being reduced to the terms of the global money market, or commodity market. Is it not preposterous to hear India and China – cultures thousands of years old, with written histories and customs more deeply rooted than anything in the West – referred to as merely 'developing nations'?[64] And yet we accept this terminology, just as once upon a time the Europeans accepted their 'duty' to save the souls of native peoples in the Americas, even if to do so meant killing most of them. Today, any culture undergoing the impact of the new media technologies will find their traditional arts and crafts swamped by mass entertainment 'product' from Hollywood or New York or London. And while we may agree that some technologies, especially some new media, will accommodate certain reforms of modernity, we must by the same token recognize that they may be fundamentally incompatible with the archaic patterns of cultures not built upon the same assumptions.

Bucky Fuller's great appeal was – and remains – his use of modern scientific and technical information to tell a different story from that of mainstream Western technology, which has too often tended mainly to bring us better guns and bigger bombs, flashier cars and cheaper junk food. Fuller tells us instead a story of humans who discover they belong to a living, responsive Universe and find ways to work in partnership with Nature – a story strikingly congruent, as we have seen, with the way of life extolled by a long succession of indigenous American visionaries. The American dream once meant something: a cooperative way of living in community with other human beings in a particular place and time. Yet the effort to universalize this 'American way of life' by modern mass media and propaganda techniques has ended by fatally betraying that vision, both at home and abroad. In the 1980s and '90s, the American dream seemed to mean little more to most people than making as much money as you can, no matter who pays. After September 11, 2001, Americans began to realize how high that price may turn out to be. It is almost a truism that today's very powerful media of communication tend to co-opt the aims, and obscure the origins, of any 'message' they convey. As soon as you start making propaganda for 'good' things, they start turning into bad things – it's in the nature of the beast, as Jacques Ellul's study of propaganda in technological societies demonstrated over three decades ago.[65] By the same token, the many 'new' scientific stories of Man in Universe, whether told by Bucky Fuller or more recently by Thomas Berry or Ilya Progogine, are prone to the fatal error of assuming that the latest discovery is the last word – that now we finally have 'got it' and all other views are obsolete. Mumford's critique of utopias applies as least as much to the new computer mandarins and electronic missionaries peddling their magical wares to 'developing' nations ('for their own good') as it did to Plato's *Republic* or *Erewhon.* However valid our concepts may be in the contexts in which they are conceived, they all too often turn toxic when extrapolated as universal values. In order to avoid doing irrevocable evil, maybe we have to stop absolutizing our own most treasured concepts of the good. A little humility goes a long way. Many a 'native' cosmology has long been telling us that Man belonged to the Earth, not the other way round, but we've been so busy broadcasting our own views that we have almost forgotten how to listen …

The upshot is that local knowledge has found its defenders, rallying around the effort to retrieve, or at least vigorously defend, all that has been eroded or marginalized by the new global information technologies. It may now be possible to use remote sensing equipment and Internet relays to put the equivalent of Bucky's Geoscope into everybody's home. (Google Earth and other ways to download local information from global positioning satellites are steps in this direction, but do you really want your neighbors peering over your backyard fence and into your windows?) The battle-lines have already been drawn between the new world information order and indigenous or traditional cultures all around us. The name of this dangerous new game is 'Jihad vs. McWorld,' as Benjamin Barber put it a decade ago, or at least it will be if we cannot find more peaceable solutions.[66] The split we examined between Fuller's global vision and Mumford's local knowledge were only two North American exemplars of a debate which has been rekindled everywhere the new technologies take hold.[67]

We humans are contrary creatures: we all seek a certain unity in our lives, but we will resist, even unto death, being reduced to anybody else's idea of unity. The hoary old problem of the One and the Many has jumped right out of the philosophy books to become the outstanding economic and political dilemma of our day, and the centerpiece of the evening news. Despite all the apparent signs of unification – e-mail, cellphones, the World Wide Web, global news media, so-called free trade, a militarized (though mainly Anglo-Saxon) New World Order, an ever-expanding European Union, etc. – we are simultaneously facing incompatibly 'distinct' societies, breakaway republics, religious fundamentalists and fanatical nationalisms everywhere. Globalization has been both relentless and superficial; the differences run too deep. One might pinpoint a European parallel to the philosophical fracture between Mumford and Fuller in the celebrated Gadamer/Habermas debates of the early 1970s, which brought to light similarly diverging vectors in the hermeneutic philosopher's archaic thrust to recover the primordial 'Word,' and the critical theorist's futuristic thrust toward an 'ideal speaking situation.'[68] Debate we have aplenty, but very little genuine dialogue – where each partner is able to assimilate and build upon at least *some* of the insights of the other.

Ambivalence about the character and direction of the technological juggernaut is scarcely news; it has vexed modern culture since at least the beginnings of the industrial revolution. In the middle of the nineteenth century, you could scarcely miss the traditionalist John Ruskin's penetrating and prophetic critique of Joseph Paxton's vast Crystal Palace – the futuristic glass and iron framework for the 1851 London Exhibition – which pioneered the daring use of metal trusses to span wide exposition spaces. He saw it as a new formalism, a rational prison for the imagination, leaving but cramped spaces for the restless creativity of those who follow:

> *The furnace and the forge shall be at your service: you shall draw out your plates of glass and beat out your bars of iron till you have encompassed us all, – if your style is of the practical kind, – with endless perspective of black skeleton and blinding square, – or if your style is to be of the ideal kind, – you shall wreathe your streets with ductile leafage, and roof them with variegated crystal – you shall put, if you will, all London under one blazing dome of many colours that shall light the clouds round it with its flashing as far as the sea. And still, I ask you, what after this? Do you suppose those imaginations of yours will ever lie down there asleep beneath the shade of your iron leafage, or within the coloured light of your enchanted dome? Not so … all the metal and glass that were ever melted have not so much weight in them as will clog the wings of one human spirit's aspiration.*[69]

You will of course see Bucky's domes as the more 'ideal' and audacious evolution of that engineering tradition, which Ruskin rhymes with Coleridge's "stately pleasure-dome" in Xanadu, while today's ubiquitous supermalls have emerged as its more 'practical' and mundane descendants, with their "endless perspective of black skeleton and blinding square." Though not geodesic, the Millennium Dome at Greenwich – centerpiece of Britain's Year 2000 celebrations, which notoriously failed to attract the expected throngs of paying visitors – invoked both Bucky's and Paxton's vast pavilions by grandiosely attempting to launch such public 'pleasure-dome' exhibitions into the twenty-first century, adding its own high-tech circus atmosphere of rock music and digital special effects under the big top. The technological society drives relentlessly forward, Marshall McLuhan used to say, all the while peering resolutely into the rearview mirror. Stepping back from these eye-popping spectacles and exhibitions, you may also discern how consciously Mumford echoes Ruskin whenever he defends the human and architectural qualities of the medieval polytechnic tradition against any and all such mechanistic attempts, in Ruskin's own words, "to invent a 'new style' … worthy of our engines and telegraphs, as expansive as steam, as sparkling as electricity." You may surmise for yourself what either thinker would make of today's formless, 'deconstructivist' junk-architecture in the highly favored and richly endowed non-style of a Rem Koolhaas.

However this may be, we today must continue to wrestle with our problem of global

The Dream of the Dome

With Ruskin's "enchanted dome" an element of the uncanny enters our story; we may call it the Dream of the Dome. It is not entirely a digression to record it here, though it does not readily 'fit' anywhere, at least not into the normal contours of experience.

We have already noticed Ruskin subsuming Paxton's Crystal Palace into a literary image drawn from Samuel Taylor Coleridge's famously fragmentary poem *Kubla Khan*, composed, its author tells us, in a dream of 1797, and only partially transcribed. The tale is well known: In an 1816 Preface to the poem, Coleridge relates that "in consequence of a slight indisposition" he had taken an anodyne, "from the effects of which he fell asleep in his chair" whilst reading in *Purchas's Pilgrimage* a line to the effect that "Here the Khan Kubla commanded a palace to be built ..." He says he dreamed a composition of some two to three hundred lines "in which all the images rose up before him as things." Upon waking he set pen to paper, but was interrupted by a visitor before he could transcribe more than the fifty-odd lines of the vivid and musical fragment which describes "a miracle of rare device/ a sunny pleasure-dome with caves of ice."

It was, I believe, Jorge Luis Borges who first called attention to the extraordinary coincidence – if the meaning of the word may be stretched so far – concerning Coleridge's dream. He tells us that twenty years after *Kubla Khan* was published, the first Western version of a Persian 'universal history' called the *Compendium of Histories* by Rashid al-Din appeared in Paris, containing this line written by a vizier of Ghazan Mahmud, a descendent of Kublai: "East of Shang-tu, Kublai Khan built a palace according to a plan that he had seen in a dream and retained in his memory." Borges justifiably asks, "How is it to be explained? ... A Mongolian emperor, in the thirteenth century, dreams a palace and builds it according to his vision; in the eighteenth century, an English poet, who could not have known that this construction was derived from a dream, dreams a poem about the palace."

He attempts various hypotheses – coincidence (improbable), an unknown text somehow known only to Coleridge (no evidence), a meeting of the emperor's soul with the poet on the psychic plane, perhaps, or maybe even an immortal or superhuman dreamer who dreams both men – but all are unsatisfying. He notes that "ruins were all that was left of Kublai Khan's palace; of the poem ... barely fifty lines were salvaged," and speculates that the dream of the dome has yet to run its course, that the full vision is yet to be realized. Borges concludes: "Perhaps an archetype not yet revealed to mankind, an eternal object ... is gradually entering the world; its first manifestation was the palace; its second, the poem." To bring his story up to date, I need record here only the bare fact that Borges published this reflection in 1951, the very period – I should no doubt say, at the very moment – that Bucky Fuller, unknown to Borges, began building his first geodesic domes, making the transition from scale-models (See Appendix A) to full sized-structures, or, if you like, from dream to realization.

In 1851, exactly one hundred years before Bucky's geodesic domes, Ruskin saw in the Crystal Palace a nightmare which for him ominously presaged the future directions of 'modern' architecture. And his prediction has, by and large, come true. But thanks to the meticulous scholarship of Borges, we are entitled to glimpse Bucky playing his part in another kind of dream altogether ... dreaming up a way to realize an 'impossible' dome dreamt long ago by an English poet, of a palace built ages earlier by the legendary Mongol Emperor of China according to a plan he himself had seen in a dream. Just whose dream (or labyrinthine nightmare) is this anyway?!

and local, futurist and archaic thinkers. We can see that this very antinomy has continued to divide entire generations of thinkers. For every McLuhan extolling the virtues of media revolutions (and comparing Bucky to Leonardo da Vinci), there will be a George Grant predicting nihilistic apocalypse. In my own generation, we have Stewart Brand, his bestselling *Whole Earth Catalogue* under his arm (inspired by Bucky, you will recall), hitching a ride on the new wave of media technologies at the MIT Media Lab, while Jerry Mander, having made his *Four Arguments for the Elimination of Television*, takes his stand with native peoples against new media as merely revamped forms of corporate-sponsored colonialism. About a decade ago, I had the good fortune to spend several months in Vancouver, where one evening's local public TV documentary about cyberpunk centered on 'post-human' predictions by Bill Gibson, the young 'neuromancer' himself; and the next evening's documentary featured ethno-botanist Wade Davis, the fellow who cracked the drug for making zombies in *The Serpent and the Rainbow*, introducing viewers to Northwest Coast Indian dance as a slow, rhythmic, ritualized alternative for people worn down by the accelerated pace of the computer-driven workaday world. In the same town, you feel impelled to visit the Science Centre, housed under a glittering geodesic dome which turns out to be far more interesting than the exhibits inside, while the renaissance of Northwest Coast Native art – which utilizes a formal vocabulary nearly as sophisticated as Fuller's, also based on the permutations of a few simple polyhedral forms – tugs at you from the other direction in the museums and galleries nearby. The same rift is polarizing succeeding generations, and we still have no middle ground. Each school of thought excommunicates the other.

Ideology is the demythicized part of your world-view; the flexible *mythos* has become inflexible *logos*.[70] Once articulated as an ideology, logic seems to dictate that you can't have it both ways: futurism is but the latest evolutionary, progressive variation on the great theme and central imaginative projection of literate, Western civilization – salvation is to be found in the 'next' world. It is a transcendent vision, always just a bit beyond reach. And archaism runs in exactly the contrary direction: salvation lies in a retrieval of the past perfect state, which probably wasn't so perfect, but certainly is past. The inspiration here is immanent, and ever more deeply buried.

We've allowed Bucky Fuller to stand for the futurist vision, and Lewis Mumford for the archaic. But this is something of a simplification; people are far too complex merely to stand for 'positions.' Mumford, too, dreamt and wrote of "World Culture" in a way which also sometimes tended to uncritically extrapolate Western paradigms. Maybe you have to really live in other cultures to see your own for what it is, and Mumford (aside from a brief stint in England) never did. As for Fuller, he went to great pains to preserve his family's retreat at Bear Island in an entirely pre-industrial condition – no electricity, no plumbing, no phones or fancy information technologies. And yet he returned to Bear Island for a month every year to 're-source' himself, as it were, and did not find drawing water from the well by hand to be incompatible with what he liked to call his 'cosmic fishing' for ideas in the ocean of intuition.

Like Emerson before him, Mumford stayed more or less centered in one place, at home on his farm in far eastern New York, but was at pains to furnish himself with a remarkably thorough grounding in world civilizations, as well as European literature, philosophy and history. Fuller moved around a lot, circumnavigating the globe many times during his long life, but he had for planet Earth the same sort of filial attachment and sentimental affection most people feel for their own little garden plot. For Fuller, the whole Earth is our local patch of Universe. Ozone depletion and the runaway greenhouse effect indicate that the stewardship of the planet he extolled is indeed long overdue, and in practice not in the least incompatible with local knowledge and indigenous customs. As ecologist Edward Goldsmith reminds us in his magnum opus *The Way*, traditional and ecological values stem from the same source, a reverence for 'the Way' things are.[71] If we permit ecology to become merely 'resource management,' it loses not only its soul – the very soul of the world, as we have seen – but all memory of those indigenous 'life-ways.'

One fairly obvious way to gauge the level of optimism or pessimism about technology in our society is to watch its science-fiction films. The new 'bad futures' we see so much of these days almost invariably have to do with getting stuck in a bad place: the near-future Los Angeles of *Blade Runner* set the tone, but a spate of films all the way back to *Escape from New York* emphasize that the places we actually dwell are very easily envisaged as prisons. This may be part of what Susan George has called "the boomerang effect," when the conditions of the so-called Third World (a two-tier society, endemic violence, decay of physical, social and moral infrastructure, etc.) begin to show up in the inner cities of America; when the beasts and horrors let loose first by colonialism, and more recently by the neo-colonialism of 'development,' come home to roost. Stephen King's kind of 'domestic' horror in his early books applies too, when the everyday objects of a technological society turn upon us and become malevolent. And did you notice the resurgence of alien abduction films all through the 1990s? People feel the society they once trusted is out of control, an anonymous and inhuman system; they feel cheated, manipulated, and treated like guinea pigs for social and economic experiments. Their lives have been hijacked by unseen agencies. Nothing makes sense anymore. Is it any wonder our collective hallucinations have strayed to the skies?

By the same token, the *Star Trek* phenomenon, now forty years old and still running strong, seems to be the only science-fiction projection which has managed to preserve an optimistic view of the future. Why? Maybe because their 'continuing mission' is always to be on the move, never to dwell permanently in one place, always to seek new frontiers where 'no one has gone before.' It is no mere curiosity that the crew of the Starship Enterprise for a time joined the astronaut exhibits in the Smithsonian Aerospace Museum, as if they were real – which in a sense they are, for an America desperate to hang on to its unbounded frontier dynamisms. There's something deep in the collective psyche at work here, something that runs like a subterranean river below all these American dreams and nightmares and technocratic fantasies of control.

Perhaps the very word 'dwelling' provides a hint ... It is well-known that in English, and indeed all the Indo-European languages, the verb *to dwell* has a rather curious etymology. We usually suppose, as Bucky Fuller did in designing his dome homes, that dwelling means something altogether positive – nesting, making a home for yourself, settling down – but these are later overlays. The word itself comes from the Anglo-Saxon *dwellan*, which means to lead or go astray, to stay (in the active sense), to hinder, wander or tarry, even to mislead or err.[72] It ultimately derives from the Aryan root *dhwel, dhul,* appearing in the Sanskrit *dhwr, dhur,* which means to mislead, deceive. One strong clue as to why 'dwelling' has such a negative connotation in the Indo-European languages may be that it is also applied to horses. When they pause and dither before jumping a fence, or when they are slow in raising their feet from the ground in stepping, horses are said to 'dwell.'

Why should this peculiar 'horse-sense' of the word be important? Because, as Sidney Pobihushchy once observed, "We are the horse-people."[73] By we, he didn't mean everybody; he meant specifically the light-skinned Indo-European races from the steppes of the Caucasus, those Indo-Aryans who long ago invaded India and pushed the darker-skinned Dravidians south; those Europeans who continued to move ever forward and outward, past the boundaries of Europe itself into New Spain and New France and New England. Of course it's not some racial quirk, as if white people can't sit still. It's those horses – the Indo-European peoples have always been on the move, leaving the graves of their forebears behind, seeking their salvation just over the horizon, 'progressing' ever forward, defining Life itself as 'evolution,' and in our own day aiming for the far planets and the stars. In its edition of 1958, the year the talkative Fuller met a silent pandit Nehru, under the entry 'Horse' the *Encyclopædia Brittanica* averred:

> *All the great early civilizations were the products of horse-owning, horse-breeding and horse-using nations and those in which the horse was either unknown on in the feral state remained sunk in savagery. No great forward movement of mankind was made without the assistance of the equine race. So consistently was this the case that the glorified figure of 'the man on horseback' became the symbol of power.*[74]

Such imperial hubris requires little comment. Incredibly, there's more to the story. As Mumford himself pointed out, the mace – a weapon specifically designed for no other purpose than to bash human heads from horseback – soon enough became, with a bit of gilding, the very sceptre wielded by the sovereign: "When Parliament is in session, a gigantic specimen lies upon the Speaker's Table."[75] Indeed, as Lynn White first clearly showed, the invention of the stirrup not only lifted the knight in full armor atop the 'great horse' bred for this purpose, but in the same stroke hoisted the entire feudal hierarchy on top of the local peasantry. Beyond this, domesticating the powerful horse begins the long quest of Western peoples to tame Nature herself and harness her inner power, a quest which has today culminated in nuclear weapons, genetic engineering and social mechanisms of technocratic control in general. People and peoples are set whirling in motion, and all the while new clothes are made to suit new climates, new inventions are concocted to fit constantly changing circumstances, and new theories, new philosophies and new sciences keep tumbling out, purportedly obsolescing and unsettling all 'settled' beliefs. And what of the local people, the native inhabitants, those lazy stay-at-homes? Obviously, they've just been "dwelling" all the while, going astray, going nowhere; surely they are in error, perplexed, misled by superstition, hindered by cumbersome traditions which have sadly retarded their 'development.' Sound familiar?

It started with horses, then it built ships, railroads, 'horseless' carriages, airplanes and rocket ships and all the rest. But where is there left to go? The vast outward thrust of the Indo-European peoples seems now to have reached – if not indeed to have grossly overstepped – its natural limits; it has become global, and circled right back in ambush upon itself. During both Gulf Wars, for instance, the US and allied Western powers of the New World Order set about systematically bombing the original sites of Western civilization in Iraq (Mesopotamia), all for the sake of fossil fuels, no doubt, but also in the name of civilization (democracy) beating back barbarism (tyranny). And yet Mesopotamia is precisely where the military-industrial complex we call '*civi*-lization' – which simply means 'city-culture' – first arose. In Chechnya, to take another hotspot for example, the local Caucasians who actually stuck around the Caucasus Mountains have found themselves at war for more than a decade with a Russia determined to join this bright and shining New World Order of conspicuous consumers. It all comes full circle. The global dynamisms of science and technology are only the latest horse we are riding; the local resistance will probably end up breeding some of their own, that is to say opening up their own website so as to assert the autonomy of their local culture over against the phony standardization of 'McWorld' (remember the flurry of faxes during the Tienamin Square standoff?). Beyond this, the global rash of debilitating computer viruses indicates that not all 'Netizens' are happy campers these days. And one day there must come a reckoning – with IT&T or Telecom or COMSAT over the phone bill, if nothing else. So the tension between globalizing technologies and local traditions goes on, adding new techniques, new torques and tensions every year.

With an historical background of this scope, the spat between Fuller and Mumford looks like a fairly minor episode. Even the glaring discrepancy between today's America – embattled land of the free market competitors – and the original American dream of cooperation between diverse peoples seems but a modest chapter of some 300 years.

Unless, of course, it turns out to be the final chapter ...

Connections – Real and Unreal

It is for me dismaying to realize that nowadays, due no doubt to plenty of high-powered and expensive publicity, Disney's Epcot Center is probably the best-known geodesic dome in the world. It's not that it's not beautiful, or that this Kaiser Geodesics' construction isn't credited to Buckminster Fuller directly that bothers me. It's that insipid Mickey Mouse waving at me from the top of the dome in the advertisements, and the idea that a geodesic dome should have become the flagship for Disney World. Now you may have charming childhood memories from Disney films, or you may be one of the debunkers (Bucky is said to have coined the word 'debunk,' by the way, in 1927) who claims that his fairytales promote sexist stereotypes or rampant consumerism or what all, but either way you will have to admit that the association with Disney World leaves Bucky and whatever vision he might have had smack in the middle of fantasy-land. It's the unreality that bothers me. (Disney World seems to be Jean Baudrillard's favorite example of what he calls 'hyper-reality,' where simulations have run amok: more real than real). I guess what it comes down to is that I just cannot dismiss the distinct impression, which has over the years congealed into something like a conviction, that Bucky Fuller was the most practical man I ever met.

Besides the Disney dome, Fuller's geometry has shown up somewhere else that places it all somehow in the world of the unreal. Science-fiction films tend to use omni-triangulated matrices whenever they want to instantly convince you that some structure or species or civilization is really 'alien.' The convention may or may not have begun with the three-eyed, three-fingered Martians in George Pal's *War of the Worlds* (1953), but threefold patterns have consistently been used since the 1950s for their weird 'otherness.' There's even an early *Star Trek* episode called "By Any Other Name" in which aliens from the Andromeda galaxy reduce everybody on the Enterprise to vector equilibria, which are left littering the floor like little urns of untouchable human remains. Now if Bucky was right – and all the indications suggest that he was – then Nature is using 60° tetrahedral coordination for all the living carbon-based systems on our planet, not to mention the metals and rock crystals as well. Apparently, science's rectilinear grid and square frame of reference has been so ingrained over the past three or four centuries that we consistently see not only 'Nature's Coordinate System' but Nature herself as something alien. Maybe it is we who have all this while been 'alienated' from our own home planet. Maybe it's time we took a look around, as Fuller himself suggested, and got to know the old home place a little better.

At least the time has surely come for us to try to evaluate Bucky Fuller's most fundamental claim: whether we have, as he put it, 'the option to make it' here on planet Earth; whether human beings were indeed 'designed for success' in this biosphere; and whether the bloodless 'design science revolution' he envisaged might help shift us from weaponry to 'livingry,' from the greedy, short-sighted profit-taking of multinational corporate 'Giants,' as Fuller dubbed them, to world-around service industries for the enhancement of human life. When all is said and done, was Bucky's claim realistic, or sheer fantasy and science-fiction?

Whatever response I might make here, the kind of answer you will accept to this question hinges almost entirely upon what you take to be real.

If what is real for you is the current set-up you tune in with the evening news on TV, then you have two choices. You may buy the whole package at face value, in which case the 'utopia or oblivion' dilemma doesn't even arise. This is just the way the world is, was, and ever shall be. Or else, option number two, if you are painfully honest with yourself you will have to shake your head and – sadly, perhaps reluctantly – accept only an answer in the negative. If the current regime of transnational corporations and international finance based on scarcity economics with its political puppet-shows and unabated environmental depredation and vicious little wars every so often to keep the arms market flourishing; if this brave 'New World Order,' along with all its ever mightier powers of illusion and collective distractions – flashy consumer commodities, trite sports spectacles, mass entertainment, celebrity scandals and all the rest – if this is what is real for you, then the answer is frankly no. If you are realistic, if you face these facts and their sobering implications as Lewis Mumford surely did, then you will have to admit that without an abrupt turnabout in basic attitudes, it is not very likely that the human species will survive its own predatory, greedy, selfish, paranoid tendencies very far into the twenty-first century. A few lucky minorities who can hang onto what they've got might have a nice run for a few years, but sooner or later the world is going to hell in a handbasket and, like it or not, we're all going with it. Entropy decrees that disorder increases; it's the way of the world. Malthus was right after all, there never will be enough to go around; so you'd better get yours while the getting is good (back to option number one), and devil take the hindmost. Moreover, there's nothing much any lone individual can do about it – except maybe to amuse ourselves in 'virtual' cyber-playpens until our day is done. Our fate is sealed by forces beyond our control, Q.E.D.

If on the other hand what is real for you is what was real for Bucky Fuller – a magnificent garden planet, beautifully designed to support all the forms of life upon it, enough air, water, soil and sunlight for everybody, all the elements of life coordinated by exquisite metaphysical principles innately apprehensible by the human mind, itself far and away the most syntropic phenomenon in the known Universe – then the panorama swiftly changes. Then an affirmative answer appears much more plausible. Then the integrity of each and every human being counts, precisely because it is intimately connected to the integrity of the Universe at large and capable of spontaneous innovation, free creativity, and the kind of playful interaction with the world around us which will bring the resourcefulness of the entire Universe to bear on local problems. In such a case, we might just surprise ourselves. If this fabric of connections is what is real, then to be realistic means to realize that despite all the overt and covert machinations of all the corporations and politicians and all the rest, it is Mother Earth and the Universe at large which support us after all, and we have been given all we need for a full and happy human life.

By contrast, the fundamental assumption of modern economics is that there is not enough to go around, and that when it comes right down to it you and I have to compete with one another to get our hands on increasingly scarce resources. Is it true? If you equate wealth with land, or with gold and other precious metals, or with fossil fuels, it would seem to be so. These things are indeed finite, and human demands upon them can only increase as human population increases. A grim prospect ... But some thinkers beg to differ, and differ radically, in their assessment of what constitutes not only wealth, but human well-being.

Buckminster Fuller, for one, defined wealth as having two components: physical energy, and 'metaphysical' know-how. There's plenty of energy coming in from the sun every day, he observed, and human know-how only increases. Thus was he able to claim that we are all 'billionaires,' and that it is high time we came into our rightful inheritance.[76] E. F. Schumacher, for another, considered fossil fuels and metals and the like to be 'planetary capital,' and insisted, like Fuller, that we must learn to live on our energy 'income' rather than by burning up our energy 'savings.' Thus was he able to claim, "There is no economic problem and, in a sense, there never has been."[77]

But modern accounting systems are rigged not to show the true cost of using up non-renewable resources, let alone the wear and tear on people as the biosphere that has so long sustained them is drastically degraded. Fuller once challenged a famous geologist to calculate what it costs nature (it takes hundreds of centuries of heat and pressure) to produce a single gallon of oil at current retail rates. The real cost worked out to roughly a million dollars a gallon. It would be much more today. At those prices, Fuller reasoned, we ought to pay people handsomely to stay home and dream up alternatives to internal combustion instead of driving their cars to work every day.

From this angle, the Earth is an economy of abundance; we need only learn how to harvest our wealth judiciously. Moreover, you and I do not in the final analysis have to steal from one another just to survive. This is good news indeed, but by now it should sound familiar: "All we humans are required to do is waste no life and be grateful daily to all life," declared the Mohawk Thanksgiving Address we examined at the outset.

"Amongst the American Indians," Fuller wrote in *And It Came to Pass – Not to Stay*, "it was the nations, not the chiefs or individuals, who controlled the land. The Indians assumed that they had only the hunting, fishing, cultivation and dwelling rights. All the land, water and sky belonged exclusively to The Great Spirit. And the same concept was held by Africans, Eskimos, and Austronesians. But it was long ago conceded in Europe and Asia that … the many individual warlords controlling the lands were the owners of all wealth." In much the same vein, Bucky reprinted in *Critical Path* an astonishing address purportedly delivered in 1851 by Chief Seattle of the Suquamish tribe near Washington's Puget Sound, his response to a federal offer to buy two million acres of land for $150,000. Well after Bucky's death, the Seattle Speech has been denounced as a hoax by scholars who unearthed evidence that it was in fact edited if not rewritten by a white Christian writer in the 1920s. Once the author is discredited, the argument implies, what the text says may also be dismissed. Bucky never knew about any of this, of course; he simply found the Seattle Speech (first presented by Dr. Glenn Olds at an Alaska Future Frontiers Conference in 1979) a profoundly moving early ecological warning and a caution about over-extending the principle of ownership. For my part, although it isn't very politically correct to say so these days, I really don't think the current brouhaha much matters. Such a message is just as real, as true and ultimately convincing, as you believe it to be. A few snippets:

> *How can you buy or sell the sky, the warmth of the land? The idea is strange to us.*
>
> *If we do not own the freshness of the air and the sparkle of the water, how can you buy them?*
>
> *Every part of this earth is sacred to my people. Every shining pine needle, every sandy shore, every mist in the dark woods, every clearing and humming insect is holy in the memory and experience of my people. The sap which courses through the trees carries the memories of the red man.*
>
> *The white man's dead forget the country of their birth when they go to walk among the stars. Our dead never forget this beautiful earth, for it is the mother of the red man. We are part of the earth and it is part of us. The perfumed flowers are our sisters; the deer, the horse, the great eagle, these are our brothers. The rocky crests, the juices in the meadows, the body heat of the pony, and man – all belong to the same family …*
>
> *We know that the white man does not understand our ways. One portion of land is the same to him as the next, for he is a stranger who comes in the night and takes from the land whatever he needs. The earth is not his mother, but his enemy, and when he has conquered it, he moves on … He treats his mother, the earth, and his brother, the sky, as things to be bought, plundered, sold like sheep or bright beads. His appetite will devour the earth and leave behind only a desert …*
>
> *This we know: the earth does not belong to man; man belongs to the earth. All things are connected …*[78]

It is easy to see why such passages appealed to Bucky, however dubious their authorship later turned out to be. A good part of the speech goes directly to mitigate the notion of property, ownership, and the extension of economic values into realms of human and natural life where they simply have no place. At the point where he cites Chief Seattle, Bucky is in the midst of one of his fables about the Great Pirates and Giants – like the fearsome man on horseback wielding a club – which eventually became limited liability colonial enterprises like the British East India Company, and are currently doing business as transnational corporations (many of which seem, as lamented earlier, altogether too fond of abusing the word 'synergistic' to describe their money-making operations). It is curious that when people read *Synergetics*, they accuse Bucky of making things too complex and technical for laypeople to understand. And then when they read *Critical Path* or *Grunch of Giants*, where he tells his tall tales of global greed and corruption ('Grunch' is Bucky's neologism for 'Gross Universal Cash Heist'), they turn around and accuse him of naive oversimplification. I wonder. Anyone who is today convinced by Susan George (*How the Other Half Dies; A Fate Worse than Debt*) or Jerry Mander (*In the Absence of the Sacred*) would be sure to recognize Bucky's fables as genuine precursors of these current, and still quite controversial, critiques of the System: "Heads We Win, Tails You Lose," as Bucky puts it bluntly in *Grunch*;[79] or, as we heard him say earlier, "You better believe it's a conspiracy."

The other resonance in such a passage goes much deeper, into Bucky's own sense of the real. It has to do with the profound intuition that humans are an integral part of a living Universe expressed by whoever it was that wrote Chief Seattle's speech, the sense of kinship between the multifoliate forms of life here on planet Earth: "All belong to the same family." "Of all his attributes," Norman Cousins wrote just after Fuller's death, "none was more compelling than his ability to transmit to others his kinship with the Universe ... I have known very few people who, after meeting Bucky, did not forever feel a sublime wonder when looking at a starlit sky."[80]

This takes us to the core of the matter. Fuller's geometry, the isotropic vector matrix and its derivatives, can all too easily seem a mathematical abstraction, a cold formalism. It isn't. It is a map of the possibility for relationships articulated between anyone or anything and everything else that exists. And this in a double sense: first, an inward confidence that the web of life on this planet already forms such a matrix: 'Nature's Coordinate System.' And secondly, an outward thrust, that this net of relationships can and must be articulated in new architectural, social and cultural forms: the 'design science revolution.' Although Fuller might at first glance appear to be merely an apologist for technological progress, the more deeply we examine his work the more the two vectors of archaism and futurism we sighted earlier appear to converge.

Every so often, there are notions which seem to galvanize entire generations, 'ideas in the air' which people will spell out in the most diverse ways, an identifiable *Zeitgeist* which is somehow shared without any direct causal agency or influence you can put your finger on. Such phenomena signal changes in the *mythos*, the underlying substrate from which all the concepts and ideas (and even competing ideologies) are born. A myth is not an idea, it is the horizon of intelligibility over against which ideas seem or do not seem to make sense. A living myth "goes without saying," in Raimon Panikkar's phrase; it seems self-evident, it is "what you believe without believing you believe in it."[81] The myth of technical progress as an indisputable good is one such horizon, currently receding. For my own post-Hiroshima generation, the emerging myth seems to be one of integral connectedness: the human, the earthly, and the sacred are all 'felt' to belong together, 'experienced' as a complex unity[82] despite all the apparent divisions and conflicts between nations, cultures, academic disciplines, and institutions which rend modern life. But a myth is constitutively susceptible to a variety of interpretations. It expresses itself in many ways. In this case, as sketched earlier, there seem to be two dramatically divergent thrusts.

On the one hand, there is an immanent understanding of this deep feeling of connectedness. Some see all the connections as *already there*, deeply buried in the geophysical and biological continuities which bind us not only to one another ("all ... the same family"), but ultimately to the whole Earth. This is the sense of the ever-underlying *anima mundi*, the maternal matrix of connections we briefly examined earlier. All we need do is stand still in our place and listen, quietly meditate and receive ... and we shall discover we already have access to the entire repertory of cosmic wisdom – built into our very genes, perhaps, or maybe just blowing in the wind (*prana, pneuma, spiritus*). Such a mentality will naturally turn to indigenous cultures in order to restore primordial connections lost in the modern, 'dis-located' worldview; or, if it stays within the precincts of modern science, it will tend to articulate a 'deep' ecology which puts the web of life ahead of mere human concerns, and the well-being of the spotted owl or some other endangered species well ahead of the livelihood of lumberjacks and their families. As an ideology, this radically cosmocentric view can tend to view human life itself as no more than a pernicious parasite upon the planet, and civilization no more than a disease, *syphilization*,[83] which threatens all other forms of life.

On the other hand, there equally is a transcendent understanding of these connections, that they are *yet to be made*. So we must wire up the planet in a communications revolution, we must build satellites and fiber optic nets and information superhighways in order to connect everybody to everybody else, in order to have all the latest information about ecology, the weather, the health of the rainforests, etc., etc. We must 'access' all the libraries from our desktop computers, tune in all the television and radio programs in all the languages, we must send e-mail messages unmediated by the powers-that-be, we must decentralize these very power structures so as to allow equity and freedom, the anarchy of life and creative spontaneity to rise up where structures of oppression have so long held sway. Such a mentality will naturally, which is to say culturally, turn to the latest technology, will wax grandiloquent about virtual realities and hypertexts and media revolutions that obsolesce everything that has ever happened until now ... or rather, until tomorrow, when Microsoft's Bill Gates, seemingly the boy with the most toys lately, completes his 'Xanadu' mansion in Seattle, a new 'high' in human dwelling with 'virtual' HDTV walls upon which will be displayed all the world's art treasures at the mere push of a button. As an ideology, this radically anthropocentric view can tend to view anything in Nature as merely raw material for the cultural revolution at hand, anything 'real' as less satisfying than the virtual and digital 'special effects' and extravagant spectacles our new technologies can produce, and planet Earth itself as no more than the testing ground for a species bound to leave it behind and fly to the stars in some vast new wave of space-colonization.

More important than such extreme views, of course, are the many efforts to bring the two vectors into harmony. Take for example the flock of high-tech television documentaries on indigenous cultures, or rainforest devastation, or fancy digital displays which render the ozone hole visible. Consider the power of "the Elder Brothers' Warning" in Alan Ereira's BBC documentary on the Kogi, which for the first time enters the villages of these pre-Columbian Tyrona people to uncover previously secret lore about Aluna, the Mother, visualized as the nine levels of a hidden matrix of connections underlying all life on Earth (ring a bell?) – while purists may still maintain they have traduced the very cultural values they seek to preserve simply by broadcasting them.[84] From this angle, post-modern theorists of the Internet avidly compare it to a 'rhizome,' the tangled rootmass of certain plants (Canada thistle, iris, ginger, trillium, etc.) which can sprout freely from any point (although critics are likely to retort that cancer might be a more apt analogy), and routinely employ Fuller's figures (usually without credit) to illustrate the latest sorts of 'link-thinking' about all the sorts of connectivity available in the 'wired' world. But many such attempts at synthesis have turned out to be exciting and surprisingly hardy hybrids which are, on balance, signs of hope. Today we have for instance the 'World Music' of native peoples available to vast new audiences on cassette and compact disk, including marvelous cross-cultural collaborations like Mickey Hart's *Planet Drum* in which at least some fusion of horizons between modern technology and indigenous traditions seems to be occurring. The burgeoning 'post-colonial' literatures of people and peoples long under the heel of the West – from Africa, from India, etc. – and yet written in Western languages (English in particular), would be still another example.

The giantism and concentrated urban populations of Fuller's later grand plans correspond directly with the media projection of the image of Bucky the Showman upon the world stage. He became a global celebrity who always had something startling to show you, something to astonish. He knew how to grab the headlines – by offering to dome over your city, for instance, or to make everybody billionaires. In 1964, *Time* featured him on its cover with his head rendered by artist Boris Artzybasheff as a geodesic dome, surrounded by many of his futuristic inventions. In 2004, the U.S. Post Office issued a commemorative 37¢ stamp based on this well-known painting – initial print-run, 60 million stamps.

In this respect, many of his later pronouncements do indeed seem to go in the direction of the tide of globalization which Rem Koolhaas has more recently been exploiting in, say, the Grand Palais at Lille, or the $3 billion Universal City expansion: more spectacle than substance. Yet it could equally be argued that Bucky's real successors are not such idiosyncratic 'geniuses' of novelty (or packaging!), but rather those rarer spirits doggedly determined to build and design ecologically fine-tuned 'livingry' without trashing the planet in the process. When *Esquire* ran a survey in the early 1970s of what was 'in' and what was 'out' in popular culture, Bucky's domes were 'in' but his ideas were already in the 'out' column. In this respect, Fuller's work as a whole has much more in common with less flashy efforts less loudly trumpeted by the mainstream media – like the low-tech, low-energy 'membrane' structures of William McDonough, a serious student of nature's own design strategies who today deplores the industrial idiom of design as "a machine not for living in, but for dying in," yet sounds undeniably akin to his mentor Bucky Fuller when he writes:

> *We must recognize that every event and manifestation of nature is 'design.' Living within the laws of nature means expressing our human intention as an interdependent species – aware and grateful that we are at the mercy of sacred forces larger than ourselves, and obeying these laws in order to honor the sacred in each other and in all things. We must come to peace with and accept our place in the natural world.*[85]

In all such multifarious efforts to fuse the traditional and the progressive, the primordial and the technological, Buckminster Fuller now appears to have been a true forerunner. If he favored computer analyses of global environmental conditions, it was because he supposed such 'unbiased' data would convince people that current industrial strategies were wasteful and inefficient, and Nature's own design strategies a viable alternative. If he was receptive to the spirituality of native peoples, it was because he felt in them something like his own sense of attunement to a living universe. Since this latter dimension of Fuller's sensibility is less well-known than all the high-tech gadgets, I would like to dwell upon it for a moment.

In a way, it's all a matter of attitude. Bucky Fuller was, as we have seen, a man of faith – faith in the unshakeable integrity of the Universe and the ability of the human mind to participate in that integrity by discovering and employing its basic principles for the betterment of human life on Earth. One of the least understood of such principles, he maintained, is *precession*. At its simplest, precession is the effect of bodies in motion on other bodies in motion. It is a kind of second-order synergy, a whole system behavior unpredictable from the behavior of two or more whole systems in relation. It is why the Earth orbits the Sun at a (90°) right-angle in response to the (180°) pull of gravity, rather than falling into the Sun. It is why a gyroscope goes off at a (90°) right-angle in response to a (180°) poke from your finger. The way Bucky interpreted it, precession may be viewed as the response of the Universe at large to movement in any local system, as well as to human initiatives undertaken in any local region. It served him as an up-to-date analogue for what a medieval mystic might very well have called the grace of God.

His argument ran somewhat as follows. Bees buzz around seeking honey, but their main contribution to life on this planet is the pollinating function they accomplish almost inadvertently for the plant-world. Humans, too, buzz around like bees, seeking money as their honey. As a (precessional) side-effect or 'fallout' from all this human busy-ness, a few really life-enhancing inventions and innovations actually do arrive, seemingly by inadvertence and happenstance, which have helped us keep our heads above water thus far. "I therefore assumed," says Fuller, "that what humanity rated as 'side effects' are nature's main effects. I adopted the precessional 'side effects' as my main objective."

So Bucky made his life into one long experiment, which he dubbed "Guinea Pig B" (for Bucky). He wanted to see, he tells us in "Self-Disciplines of Buckminster Fuller," whether the principle of precession, a wonderment in the physical domain, might have socio-economic side effects.[86] If he devoted his efforts to discovering and working with Nature's principles for the improvement of human life, he assumed that the precessional 'side effect' would be that the Universe would support him and sustain his efforts materially. Bucky sought to fulfill Emerson's ideal of 'self-reliance' by rendering himself, so to speak, entirely 'Universe-reliant.' So he jettisoned the oppressive concept of 'earning a living' and all of what he called the 'honey-money bumbling' that went with it, reasoning that:

> *Since nature was clearly intent on making humans successful in support of the integrity of eternally regenerative Universe, it seemed clear that if I undertook ever more favorable physical-environment-producing artifact developments that in fact did improve the chances of all humanity's successful development, it was quite possible that nature would support my efforts … I noted that nature did not require hydrogen to 'earn a living' before allowing hydrogen to behave in the unique manner in which it does. Nature does not require that any of its intercomplementing members 'earn a living' …*

> *I went on to reason that since economic machinery and logistics consist of bodies in motion, since precession governs the interbehaviors of all bodies in motion, and since human bodies are usually in motion, precession must govern all socioeconomic behavors ...*
>
> *I assumed that nature would 'evaluate' my work as I went along. If I was doing what nature wanted done, and if I was doing it in promising ways, permitted by nature's principles, I would find my work being economically sustained – and vice-versa, in which latter negative case I must quickly cease doing what I had been doing and seek logically alternative courses until I found the new course that nature signified her approval of by providing for its physical support.*

One is struck by Fuller's confidence that Nature would correct and inflect his initiatives, a notion he extrapolated in the essay "Mistake Mystique."[87] There he reflects that humans learn mostly the way kids do, by trial and error. You can tell children to do this or not to do that all day long, but it merely increases their determination to find out for themselves. Any parent knows that kids have to make their own mistakes.

What Bucky observed was that our schools try to cut right across the grain of this natural learning: only correct answers are rewarded, and mistakes penalized. Yet if you never make mistakes, Fuller reasoned, you never learn anything new. It follows that our schools tend to produce people who follow instructions correctly, but who are wary of trying new things, thinking for themselves, or taking any risks. Is it any wonder that our arts are tendentious, that we lag behind in technological innovation, or that we hang onto bad laws and disastrous social policies until it's too late? Everybody is deathly afraid of making mistakes, or admitting them. Fuller, himself a great innovator, recommended flatly reversing educational priorities: reward students for the number of their own errors they discover and remedy, rather than for merely parroting 'correct' answers.

It could catch on. Imagine, say, advertisers apologizing for all those lies, governments freely admitting the error of their ways, corporations confessing to price-gouging, industries taking responsibility for pollution, and so forth. We might very soon be living in a different sort of world altogether. Or am I entirely mistaken? The passage in *Critical Path* continues:

> *Wherefore, I concluded that I would be informed by nature if I proceeeded in the following manner:*
>
> *(A) commited myself, my wife, and our infant daughter directly to the design, production, and demonstration of artifact accomodation of the most evident but as-yet-unattended-to human-environment-advantaging evolutionary tasks, and*
>
> *(B) paid no attention to 'earning a living' in humanity's established system, yet*
>
> *(C) found my family's and my own life's needs being unsolicitedly provided for by seemingly pure happenstance and always only 'in the nick of time,' and*
>
> *(D) being provided for 'only coincidentally,' yet found*
>
> *(E) that this only 'coincidentally,' unbudgetable, yet realistic support persisted, and did so*
>
> *(F) only so long as I continued spontaneously to commit myself unreservedly to the task of developing relevant artifacts, and if I*

(G) never tried to persuade humanity to alter its customs and viewpoints and never asked anyone to listen to me and spoke informatively to others only when they asked me so to do, and if I

(H) never undertook competitively to produce artifacts others were developing, and attended only to that which no others attended

then *I could tentatively conclude that my two assumptions were valid:*

(1) that nature might economically sustain human activity that served directly in the 'mainstream' realization of essential cosmic regeneration, which had hitherto been accomplished only through seeming 'right-angled' side effects of the chromosomically focused biological creatures; and

(2) that the generalized law of precessional behaviors does govern socioeconomic behaviors as do also the generalized laws of acceleration and ephemeralization.[88]

In short, wonder of wonders, it worked! Bucky says he adopted this course of action in 1927 (though some of the reasoning sounds *a posteriori*), and he lived and worked another fifty years without falling off the edge of the world (he would object to such a flat-earth anachronism). So Bucky offers us his own life as an example, a rather remarkably successful proof-by-experiment with "Guinea Pig B" that if what you do supports the principles of the Universe and the betterment of humankind, you can rely on the Universe to support you. Do you believe it? Probably not; it's not the sort of thing we take on faith in these days of tight money, dog-eat-dog, high unemployment, and rampant homelessness. Yet the message has been around for some time: "Blessed are the unemployed, for they shall deploy themselves ..."[89] That famous unemployed carpenter from Nazareth never said these words (the 13th Beatitude?!), but he might have. He did, after all, bid his disciples to "Consider the birds of the air. They neither sow nor reap, and yet our Father in Heaven feeds them."(Mt. 6:26) For 2,000 years, Western civilization as a whole has steadfastly refused to consider those birds – some notable monastic and utopian and counter-culture experiments along these lines notwithstanding. Will you try it? Not very likely, I suspect. Yet the only way to find out whether it's true is to turn one's entire life to the task, as Bucky did, or indeed as Gandhi did in his own 'experiments with truth.' To rely utterly on the matrix of creation itself, ignoring mass opinion and institutional structures and all the carrots and sticks pulling and pushing us in exactly the opposite direction ... this calls for a radical transformation of attitudes, and one so at odds with the conventional wisdom of our day that without a few such exemplary lives lived wholly for others, it might well seem out of reach entirely, if not just incomprehensible. I still insist that Bucky Fuller was just about the most practical person you'd ever want to meet. Maybe more so than most of us are willing, or able, or courageous enough to be.

Such considerations bring us full circle to our final point – which is also, as we shall see, a point of origin.

A Lower Deep

Brahma

If the red slayer thinks he slays,
Or if the slain think he is slain,
They know not well the subtle ways
I keep, and pass, and turn again.

Far or forgot to me is near;
Shadow and sunlight are the same;
The vanished gods to me appear;
And one to me are shame and fame.

They reckon ill who leave me out;
When me they fly, I am the wings;
I am the doubter and the doubt,
And I the hymn the Brahmin sings.

The strong gods pine for my abode,
And pine in vain the sacred Seven;
But thou, meek lover of the good!
Find me, and turn thy back on heaven.[90]

Ralph Waldo Emerson

Reflection, or re-*flexion,* means bending back to the source … So far, we've gone from Indians to India and back again, by great circle routes of sea and air and spirit. Now all these horizons converge. We've not been able to talk solely about Buckminster Fuller without also considering Emerson and the Iroquois and Mumford and mandalas and modern media and all the rest: "All things are connected," said the pseudo-Seattle. This radical connectedness and interdependence is the key intuition … so that any movement here is reciprocated by the Universe; so that whenever you take a step upon the Earth, as Bucky liked to point out, she pushes back ever so gently; and so that, sadly, restricting your focus to improvement in one area (high-tech medicine, for instance) usually entails deleterious effects elsewhere (general health and well-being declines). This web, Fuller sought to demonstrate, is the one thing you can really rely on, and it is still palpable under all the encrustrations of economics and politics. With that protractor of a mind that was his, Bucky stood inside this web and actually measured the angles where the Pseudo-Dionysius (another even more spectacular literary hoax, by the way – the original 'pious fraud') saw but the wings of angels.

Yet the two visions are quite similar … except that, as sketched earlier, the celestial hierarchy of Dionysius (three sets of three intelligences mirroring the Trinity; as they do in Dante's *Paradiso*) has been as it were plunged tip-down into the very atomic and molecular heart of matter. It is both the buried *arché* of creation, and the future utopia yet to be achieved – or rather it is neither Arcadia nor Utopia, but abides as a kind of underlying presence in the present. Immanance is not just transcendence inverted; the spatial analogy is misleading. Transcendence is absolute difference; immanence is qualified identity. It is the real presence of the Whole in every part, and 'the impress,' as Smuts phrased it, of the Whole upon its every part. And there is more: "True immanence," says R. Panikkar somewhere, "is always where transcendence is experienced." The 'turning point' (apologies to Fritjof Capra) is NOW: neither nostalgically swooning over paradises lost in the past, nor feverishly hankering after future utopias that never come, but attunement here and now to this vital matrix of connectedness. It is the intuition of the Ground of Being, notably more feminine (*mater: matrix*) and nurturing than the *Ab*-solute (meaning *un*-related) transcendence of European Christianity.

The tradition Buckminster Fuller inherited was already to some extent a cross-cultural hybrid; a vital fusion of East and West had begun to take root half a century before he was born. The Hindu "Brahma" of Emerson is not the God Brahma, but the Brahman of the *Upanishads* (with obvious reference to Kriśna's speech to Arjuna in Book XI of the *Bhagavad Gita* at the start). This is the Brahman that is *atman*, the Self at the core of every 'self'; in Emerson's vocabulary, the Over-Soul resident in every soul. The fascination of the so-called New England 'transcendentalists' with the Hindu classics is well-known; the *Gita*, for example, was as familiar to them as it was to the hippies of the 1960s. Margaret Fuller, Bucky's great-aunt as well as America's first feminist author, you will recall, published most of the reflections of Emerson and his 'circle' in *The Dial*, such notions paralleling her own fascination with the organic conception of life she found in translating Goethe: *Life Within and Life Without*, as she would put it. For his part, the woodsy Henry David Thoreau was particularly enamored of the interrogatory creation story in *Ṛg Veda* X, 129:

Who really knows? Who can presume to tell it?
Whence was it born? Whence issued this creation?
Even the Gods came after its emergence.
Then who can tell from whence it came to be?

That out of which creation has arisen,
Whether it held firm or it did not,
He who surveys it in the highest heaven,
He surely knows – or maybe He does not![91]

In this tenth 'Mandala' of the oldest human scripture, we witness something quite unfamiliar to European philosophy: a deliberate suspension of 'causal' thinking. Indeed, the question of whether a causal agency can be established, a 'prime mover' named, may well reveal a limit which the human mind, itself a creative 'moment' in that creation, cannot intelligibly transgress. Here we find ourselves in a very different mental and spiritual climate from the Abrahamic traditions. As today's New England 'geologian' Thomas Berry would remind us, landforms and bio-regions have a direct and sometimes overwhelming bearing on spirituality.[92] The three great monotheisms were and to some extent remain 'desert' religions; Nature is for them no more than a barren wasteland – it is a hard, cruel, flawed, and ultimately sinful world. One's only hope for salvation lay in transcending the 'merely' natural altogether in order to enter a supernatural Kingdom of Heaven. (The exceptions tend to be ecstatic sects like the Shakers, whose Kingdom had already come, or the many Sufi orders of Islam; in point of fact, Anwar Dil's conversations with Fuller in *Humans in Universe* seem to invite Bucky to make himself at home in that religious universe as well).[93] Yet although Emerson came from a long line of New England Protestant preachers, as did Fuller for that matter, there is no hint in his work of any such puritanical disdain for things natural. To the contrary. Emerson's own first little book, *Nature*, takes nothing less than the entire created order as the verdant 'symbol of Spirit.'[94] His first full collection of *Essays* time and again strikes this distinctively American tone of spirit. Here human creativity in art, music, literature or scientific invention is not conceived on the European model as an agonistic struggle with God – Faust; Prometheus; Jacob wrestling with the angel – a 'counter-creation,' following George Steiner's convincing description of European art and artists, so as to be 'like' God.[95] (Is this not also Victor Frankenstein's Romantic temptation?) Given Europe's long and troubled Judæo-Christian history, asserting one's 'difference' from a transcendent God – Itself defined by Its absolute 'difference' from everything earthly and natural – might well come to seem the most 'godly' thing for a human being to do. But this is not, surely, the American vision, not the consistent effort to *co-operate* with the divine 'Spirit' in Nature which we must by now recognize as the specifically 'New World' ingredient in the spirituality these American visionaries shared. In his 'Compensation,' for instance, Emerson outlined what an indologist might recognize as his own homespun theory of instant karma:

> *The world globes itself in a drop of dew. The microscope cannot find the animalcule which is less perfect for being little ... So do we put our life into every act. The true doctrine of omnipresence is, that God reappears with all his parts in every moss and cobweb. The value of the universe contrives to throw itself into every point. If the good is there, so is the evil; if the affinity, so the repulsion; if the force, so the limitation.*
>
> *Thus is the universe alive. All things are moral. That soul, which within us is a sentiment, outside of us is a law. We feel its inspiration; out there in history we can see its fatal strength. 'It is in the world, and the world was made by it.' Justice is not postponed. A perfect equity adjusts its balance in all parts of life.*[96]

With such intimations of a moral Universe at the roots of New England spirituality, the phenomena Fuller described as the socio-economic repercussions of precession don't sound so remote anymore. Humans are not to see themselves at odds with creation, as in European Christianity, but in tune with all Universe. Even in Fuller's fulminations against 'earning a living,' one cannot but hear echoes of Emerson's 'Spiritual Laws':

> *Why need you choose so painfully your place, and occupation, and associates, and modes of action, and of entertainment? ... For you there is a reality, a fit place and congenial duties. Place yourself in the middle of the stream of power and wisdom which animates all whom it floats, and you are without effort impelled to truth, to right, and a perfect contentment ...*
>
> *I say,* do not choose ... *Each man has his own vocation. The talent is the call. There is one direction in which all space is open to him. He has faculties silently inviting him thither to endless exertion. He is like a ship in a river; he runs against obstructions on every side but one; on that side all obstruction is taken away, and he sweeps serenely over a deepening channel into an infinite sea. ... He inclines to do something which is easy to him, and good when it is done, but which no other man can do. He has no rival. For the more truly he consults his own powers, the more difference will his work exhibit from the work of any*

other. ... By doing his work, he makes felt the need which he can supply, and creates the taste by which he is enjoyed. By doing his own work, he unfolds himself ... The common experience is that man fits himself as well as he can to the customary details of that work or trade he falls into, and tends it as a dog turns a spit. Then is he a part of the machine he moves; the man is lost. Until he can manage to communicate himself to others in his full stature and proportion, he does not yet find his vocation.[97]

This, then, may be the double legacy of Buckminster Fuller – it is all the words and works he left us to ponder, of course, but it is equally *all that was left him* by his own New England forebears. Maybe we are too much concerned with what one individual can achieve, perhaps understandably when that individual's achievements stand out as strongly as Fuller's do in the midst of the rubble and ruins left by runaway 'progress' in the twentieth century. Still, we miss something of Fuller if we do not see him taking his proper place in this main stream of American visionaries. By the same token, our literary analyses are all too preoccupied with 'influences.' It is an instance of causal thinking running riot beyond its ken. Did Fuller read Emerson's *Essays*? I cannot imagine him *not* reading them, sometime in his youth perhaps, but direct references are wanting (aside from one or two rare occasions when he compares Emerson unfavorably to his great-aunt Margaret).[98] It was Mumford, not Fuller, who would claim the mantle of Emerson. And it was Hugh Kenner, not Fuller, who first spoke of the Emersonian streak in Fuller's writing.[99] Yet this business of 'influences' misses what I take to be the main point both Emerson and Fuller were trying to make: the connections are there, and very real, even though they do not readily fit into our customary causal frames of reference (customary since Aristotle, anyway). The matrix to which these thinkers try to direct our attention is not causal, but it is intensely creative. One author does not 'cause' another to write. Clairvoyants don't 'cause' the matters they 'see,' or foresee. Humans don't 'cause' the Universe to respond to them, or the Good Lord to answer their prayers. This is but a connecting matrix – like Indra's Net of Gems in the Buddhist *Flower Garland Sutra*: a web of jewels in which each pearl perfectly reflects all the others (also a precise description, you may have noticed, of an isotropic vector matrix). Causality is a logical category; creation – whether envisioned as a once-upon-a-time act of God, or as an ever-present matrix of living connections – creation is a myth.

In the century after Emerson, and quite consciously so since the 1960s, modern science has begun to come round to this intuition that along with, or besides, or above, or indeed *beneath* causality there is something *more* ... Carl Jung and Wolfgang Pauli called it 'synchronicity,' and sought to demonstrate that certain instances of coincidence like telepathy and clairvoyance were, in a phrase, no coincidence.[100] But such connections are not restricted to temporal conjunctions alone. The cohering principle is comprehensive of space as well as time. Why doesn't organic life on Earth just disintegrate? What holds entropy at bay? As we have observed Fuller intently observing, the world seems to be held together not primarily by cause-and-effect mechanisms – mainly the mechanisms of force, which tend to increase entropy, randomness and disorder – but by a remarkably resilient and reliable fabric of connections, a wholly *acausal* matrix of minutely articulated relationships, none of which may be ignored or neglected with impunity.

Arthur Koestler spent a good many years tracking evidence for this acausal matrix in several disciplines, a search of which he wrote at length in *The Roots of Coincidence* and *Janus*. Koestler stressed that while science knows how to break up nature's wholes (atomic and cellular nuclei, for example), it is only on the verge of learning to detect the integrative principle which holds these realities together. To illustrate this non-causal cohering principle, he assembled a formidable battery of evidence and authorities:

> *In the present theory, the 'order from disorder' principle is represented by the integrative tendency. We have seen that this principle can be traced all the way back to the Pythagoreans. After its temporary eclipse during the reign of reductionist orthodoxies in physics and biology, it is once more gaining ascendancy in more sophisticated versions. I have mentioned the related concepts of Schrödinger's negentropy, Szent Györgyi's syntropy, Bergson's elan vital, etc.; one might add to the list the German biologist Woltereck who coined the term 'anamorphosis' … for Nature's tendency to create new forms of life, and also L.L. Whyte's 'morphic principle,' or 'the fundamental principle of the development of pattern.' What all these theories have in common is that they regard the morphic, or formative, or syntropic tendency, Nature's striving to create order out of disorder, cosmos out of chaos, as ultimate and irreducible principles beyond causation.*[101]

In this 'kinship group' of maverick scientists, Fuller's reflections on the intrinsic connectedness of 'Nature's Co-ordinate System' would seem as much at home as they do in the 'family' of New England transcendentalists.

Once and for all, however, this term 'transcendentalism' badly needs to be qualified. Isn't the shortcoming of Delbanco's *Real American Dream* that he looks only to the Puritans for his 'native' American dream, and fails to qualify their starkly 'transcendent' God with any sense of the immanent 'being here now' that one cannot miss in Native American traditions, or later in Emerson? Originally borrowed trom the German idealism Carlyle happened to be promoting at the time in England, the term 'transcendentalism,' as used in America, draws only indirectly on the contemporary idealist philosophies of Kant, Fichte, Schelling, or Hegel (though Emerson was quite conversant with all of them).[102] Stanley Cavell, who knows Kant better than most of us ever will, maintains there is a strong connection, although of course Emerson responds to Kant in his own way.[103] To my mind, Goethe's organicism was probably the closest continental cognate to the spirit of American transcendentalism – Margaret Fuller, fluent in German, saw this early on; Emerson labored for years to read the *Gesämmelte Werke* in the original – but neither Goethe nor his New England readers were much given to dilating upon the transcendental monotheism of European Christianity, either Catholic or Protestant. Hence perhaps the appeal of Emerson to Nietzsche, for instance, rather than to clergy in his own neighborhood.

Outside of certain mystical schools, the God of European Christianity has always stood for what R. Panikkar calls a 'transcendent transcendence.'[104] The numinous is, in Rudolf Otto's renowned phrase, 'wholly Other:' distinct, different from everything else in our experience.[105] God is the ultimate 'individual.' It is also predominately 'imaged' as a masculine Supreme Being: the Father of Lights. Despite the mellower tones of the *Abba*, or a St. Francis, its attributes are mainly Force, Power and Glory – emphatically *not* the image brought forth time and again by New Englanders from Emerson to Fuller. Theirs is an 'immanent transcendence,' a feminine Nature, an all-embracing Ground of Being.[106] Not altogether surprisingly, this image seems to accord less directly with the

Jewish and Christian Yahweh than it does with Brahman of the Indic traditions. (As Emerson saw, the intuition is in point of fact not incompatible with the Christian God, but you have to go through the *Logos*, the incarnate Word, even to speak of it, let alone to 'connect' transcendence to this world: "It is in the world and the world was made by it.") Says Panikkar, "*Brahman* is equally transcendent, though not because it is distant, different and above, but precisely because it is below, common, the mere condition for being ... It is the *matrix*, the *yoni* (womb), more like a mother nurturing from below than a command from above. It does not lead but sustains."[107] In this sense, we might begin to hear Fuller's unusual meditations on the Lord's Prayer in *Intuition* as attempts to sound notes of both transcendence and immanence in discordant concord:

Oh god, our father –
our furtherer
our evolutionary integrity unfolder ...
omniexperience is your identity.[108]

So Fuller and his fellow New Englanders (not to mention some at least of the neighboring native pseudo-'Indians') would seem to be 'immanent' transcendentalists rather than 'pure' transcendentalists; a turn of phrase which may help us leave behind the dichotomy between 'archaists' and 'futurists' that has so vexed twentieth-century thinkers. In the final analysis, it may be monotheism which turns out to be the greater bugaboo here – greater even than gender. Only a crippled and myopic monotheism would try to force us to choose between Father and Mother, the Nature of the Divine and the Divine in Nature: "God without form is my Father," exclaimed Kabir, the sage fifteenth-century Muslim weaver revered today by Muslims and Hindus alike: "God with form is my Mother."[109]

Speculative? Yes, precisely. We speculate because we are a *speculum*, as the ancients used to say, a mirror of the Whole. We are the 'speculative' dimension of the entire Reality; the microcosm mirroring that macrocosm. And as long as we are speculating on the legacy Bucky both inherited and passed on, we might try listening to linguistic echoes as well as looking for visual, geometrical and structural analogies.

Western liberal philosophy has centered on individual liberties and human rights, and this is its great contribution.[110] Many traditional cultures – Indian and Indian alike, so to speak – have been built up instead around strong collective notions of duty and social responsibility. The contrast comes to a head in contemporary debates about whether Western ideas of human rights can be universalized, for instance, or whether Western attitudes toward the Earth are inherently irresponsible. A great deal might well hinge upon whether one perceives the world as a battlefield or as a living matrix of connections.

It could be said that both Hindu and Buddhist conceptions of the world revolve around differing interpretations of a single powerful root-word, which has also spread its branches into all the Indo-European languages. That root is DHṚ, pronounced *dhri* (with a very short 'i' at the end). At its simplest, the Sanskrit root *dhṛ* means to hold, maintain, keep together.[111] Most importantly for our purposes, *dhṛ* is the root from which the keyword Dharma derives. Dharma is conventionally translated into English as 'duty,' but there is much more to it. Here is Panikkar on the extraordinary complex of meanings encompassed by this one word:

> *As is well known, the meaning of the word Dharma is multivocal: besides element, data, quality and origination, it means law, norm of conduct, character of things, right, truth, ritual, morality, justice, righteousness, religion, destiny, etc. Etymologically and existentially ... Dharma is that which maintains, gives cohesion and thus strength to any given thing, to reality, and ultimately to the three worlds (*triloka*).*
>
> *Justice keeps human relations together; morality keeps one in harmony with oneself; law is the binding principle for human relations; religion is that which maintains the universe in existence; destiny is what links us with our future; truth is the internal cohesion of a thing; a quality is what pervades a thing with an homogenous character; an element is the minimum consistent particle, spiritual or material; and the like.*[112]

To make a long and very beautiful story all too short: interpreted as it were 'positively' in the Hindu tradition(s), Dharma is *that which holds the world together*. Interpreted as it were 'negatively,' e.g., by the Buddhist Nagarjuna, all the dharmas are void, *śunya*, empty, which is to say wide open, clear – a notion scarcely opaque to any observer of Bucky Fuller's synergetic structures.[113] The Buddhist intuition of the ultimate structure of things is therefore a 'transcendent immanence,' a deliberate inversion of the Hindu terms. Remarkably, the celebrated European Buddhist scholar S. T. Stcherbatsky, later killed in a Nazi concentration camp, actually made the leap to Fuller's terminology during the 1930s in his *Central Concept of Buddhism*, when he dismissed inadequate alternatives and chose to define dharmas as 'synergies,' leaning on the doctrine of *pratityasamutpada*, the radical relativity of all things.[114]

What is of deep interest here to a student of Fuller's work is the way the root *dhṛ* lies just below the surface of our own language. It is this *etym*, or *radix*, this very root which underlies **tru**th and **tr**ust and **tr**iangles and **tr**usses alike: you could say, "Trust the truth of the trees and trusses," and be saying almost nothing but *dhṛ* over and over again. The same root makes a family of the words **dr**aw and ins**tr**uct and cons**tr**uct. The words **der**iving from *dhṛ* always indicate pulling, **str**etching, **dr**awing out; in other words, **ten**sion rather than compression. And if you **tr**y listening to these linguistic echoes, you begin to realize that Fuller's at**ten**tion to **ten**sional s**tr**uctures and the intertrans**form**ability of all **forms** comes into focus through this linguistic ma**tr**ix of connected meanings. The same root binds together all the **tr**iangles and te**tr**ahedra by which Fuller main**tain**ed the entire physical cosmos is sus**tain**ed.

In fact, it is a perfectly extraordinary feature of Fuller's idiosyncratic use of language. He set out in 1927 to speak only in words he had made his own, but he did not own even those words, nor did he control them as much as he might have supposed. He was, after all, in the midst of a lifelong effort to articulate a range of notions which do in fact tie in closely to this particular etymological root. Some of the most challenging and innovative elements of his personal vocabulary consist almost entirely of *dhṛ* words, which he defined for himself in his own way. (An example, chosen practically at random: "All **str**uctures, properly understood, from the solar system to the atom, are **ten**segrity **str**uctures. Universe is omni**ten**sional integrity.")[115] What Fuller sought to articulate was nothing less than *that by which the whole coheres*, the generalized principles by virtue of which the entire Universe holds together – a notion with which the Indic traditions are perfectly at home under the heading of Dharma, imaged either as emptiness (Buddhism) or wholeness (Hinduism). Even Fuller's idea of relying on precession rather than 'earning' the life you were freely given at birth seems to have been anticipated in the great Hindu epic, the *Mahabharata*. In so many words:

If Dharma is protected, Dharma protects.
If Dharma is destroyed, Dharma destroys.[116]

How do we 'explain' this conjunction of Fuller's peculiar way of speaking with one of the eldest and most venerated linguistic roots, the very source from which entire religious and cultural traditions have drawn sustenance? We don't, we can't … but we can and we do affirm that such a connection exists, and that it amounts to more than sheer coincidence!

Maybe wherever people attune themselves to the rhythms of Nature, the same geometric forms and figures of speech Fuller spent his life articulating in modern and Western ways begin to reveal themselves. Similarly, when people try to put themselves above or outside Nature – to isolate the detail in the frame without regard for the whole picture; to put Nature in a box, so to speak – the all-too-familiar ugly patterns of environmental degradation and human misery also begin to re-assert themselves.

If nothing else, Bucky tried to show us another way. As he put it in one of his final scraps of writing: "In any act of regeneration, God, Man and Universe are spontaneous co-operators, i.e., synergists. Only the whole coheres – ergo only the whole exists. There are no parts."[117]

Appendix A

Unfolding WHOLES: A Synergetics Primer

by Scott Eastham & John Blackman

*Bucky at Black Mountain College with the first great circle models (WHOLES), 1948.
(Estate of R. Buckminster Fuller.)*

The WHOLE Story

Now for the fun part. Back in the late 1940s, when Buckminster Fuller was cooking up the first geodesic domes at Black Mountain College, he developed a series of spherical figures which elegantly model the principles of his synergetic geometry. They also served as prototypes for those first domes.

Some years ago, we two and some friends – most notably Tom Parker and Michael Connolly – started constructing these figures from specifications first published in Fuller's *Synergetics* (1975). After producing the models in clear and reflective (polyester) media, we showed them to Fuller, who pronounced them "Beautifully done," and gave us his blessing – literally: "You have my blessing" – to christen the full spectrum of seven models 'WHOLES' (pun intended), and even to market them under that name.

To distinguish WHOLES from geodesic domes and tensegrity spheres, WHOLES are *great circle models*. Only seven such figures can be constructed so as to keep the great circles intact. They make abundantly evident a dimension of Fuller's geometry – what happens at the center?! – not otherwise easily modelable.

Since the demand for WHOLES in the late 1970s quickly outstripped our ability to produce them by hand, we set about finding ways for more people to put them together more easily. A lot of little cotter and bobby pins were bent along the way. Soon we had to have dies cut to stamp out the many disks. For a couple of years, we marketed WHOLES kits of increasing precision and sophistication, mainly in Northern California where in those days we could get away with calling ourselves "The Whole Works." We received reports of WHOLES flying (well, hanging by a thread) in Europe and Japan, and even within sight of Mt. Everest in Katmandu. As late as 1999, one was spotted catching the sunlight in the Temple of the Azure Cloud atop Mt. Taishan (Confucius' Sky Mountain) deep in old China.

We sought for years to bring out a brief alternative geometry textbook which would include pre-scored disks and pins to make 'unfolding' the WHOLES a snap. But alas, publishers have shied away from such 'four-dimensional' illustrations, which would require breaking the 'frame' of paper books. It's not easy to produce a round thing in a square world ... Here we present you mainly the specifications for the figures, and some tips on construction, with Fuller's explicit permission. We also include a few of the scientific corroborations now available for Fuller's claim that Nature is indeed using the isotropic vector matrix and these elementary figures derived from it for her most basic atomic, molecular, crystalline, metallic and organic structures.

One of the tenets of synergetics is that the straight lines and cubical coordinates of conventional science and technology are simply inadequate models for Nature's spherical and cyclic patterns of growth. Consider galaxies, stars, planets, trees, flowers, rock crystals, molecules, atomic nuclei and so forth ... Nowhere do we find boxes, or frames,

or straight lines, but everywhere only wholes within wholes within wholes. This primer is intended to show how life's own spherical geometry can finally replace the mechanistic abstractions of plane geometry, and spherical polyhedra – WHOLES – supplant 'boxes' and 'frames' as the more accurate models of Nature's processes.

Furthermore, as this book (*American Dreamer*) goes some distance toward demonstrating, such patterns bear a striking resemblance to the age-old iconographies and sacred geometries which proliferate with astonishing vitality in practically every human culture. Whatever the *mandala* is said to mean in the diverse traditions that have cultivated this art – an orientation toward the WHOLE of reality, perhaps? – it may well be happening all over again in Fuller's 'geometry of thinking.'

The WHOLES, then, are first of all one array of the supremely effective heuristic devices Fuller invented and used for teaching synergetics. They also seem to be a key to existing symbolic systems, rendering the implicit connections of many traditional iconographies to natural patterns almost transparent. And they may turn out to be themselves a new medium, a vehicle for artistic endeavor through which such systems of meaning and value may yet find all sorts of further applications. That part, gentle reader, will be up to you.

For now, a primer of synergetics. And our wish that 'unfolding WHOLES' will be as much fun for you as it has been for us.

SE & JB

SOMEWHOLESOMEWHOLES

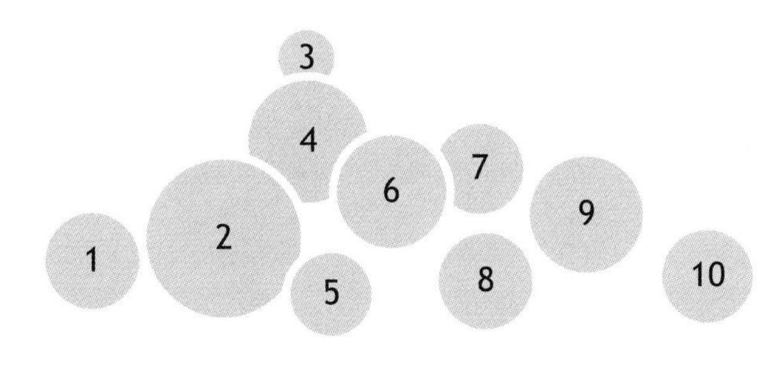

1 Spherical tetrahedron
2 Icosa
3 Vector equilibrium
4 Vector equilibrium
5 Icosa
6 Rhombic triacontahedron
7 10 great-circle icosa (Buckyball)
8 Spherical octahedron
9 Frequency vector equilibrium
10 Rhombic dodecahedron

Synergetics Primer

Synergy, from *syn-ergein* (Greek: working together, i.e., *co*-operation)

Synergy: Behavior of whole systems unpredicted by the behavior of their parts taken separately.

Corollary: Once you start dealing with the known behavior of the whole and the known behavior of some of the parts, you will quite possibly be able to discover the unknown parts.

Synergetics: The exploratory strategy of starting with the whole.[1]

1. Triangle

Energy creates pattern.
Ezra Pound

To make any headway in simplifying the structural insights of *Synergetics,* we must try to think the way Bucky Fuller himself does. So we follow Bucky's reasoning back to basics: There are no 'things' in all Universe, he points out, only events. We'll start with this ...

What we experience as 'things' are nothing but the interpatternings of relationship between events. Events are the dynamic transactions and transformations of energy.

In other words, 'things' don't just sit there: They *happen.*

Now energy moves. It takes one direction or another. It articulates certain recognizable patterns. (¶223.80) It has *shape.*

The primordial shape of any energy event is threefold: action, reaction, and resultant. (¶537.15) Its vector diagram discloses a triangle, or more precisely, a tendency to *tri*-angulate.

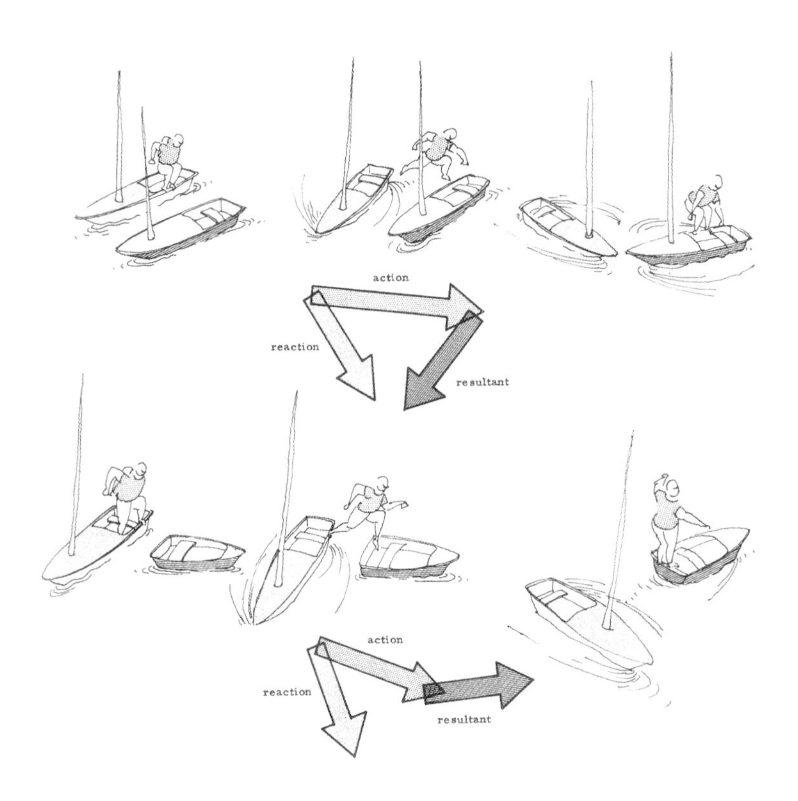

Energy, Fuller notes, always and only coheres with minimum effort and maximum efficiency: "Triangle is structure." Nothing more nor less than the triangle can make this claim.

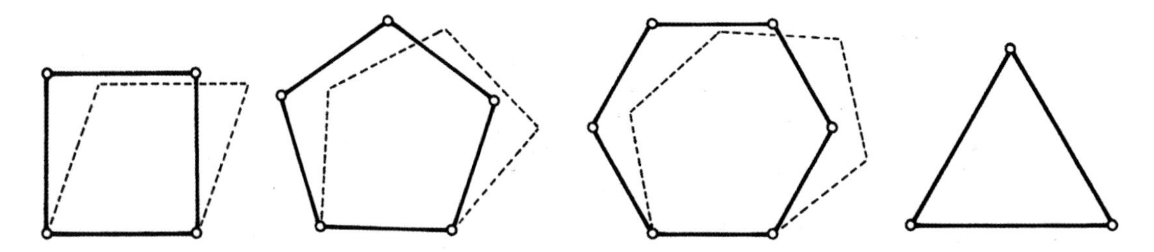

By structure, we mean a self-stabilizing pattern. The triangle is the only self-stabilizing polygon. By structure, we mean omnitriangulated. The triangle is the only structure ... Only triangularly structured patterns are regenerative patterns. Triangular structuring is pattern integrity itself. This is what we mean by structure. (¶610.01-03)

Synergetics begins with the understanding that relations are more real, and more important, than whatever they relate. You cannot encompass any relationship from only a single one of its 'perspectives' (angles). Just so, triangulation in nature is never the closed, abstract, static entity that Euclid and his schoolmarm minions would have us believe. The way of triangulation is to spiral. That uncanny third angle (and any of its angles can be the third) is really always open, even if ever so slightly – to other events, other configurations and constellations of energy. There is no depth dimension until two events come together – and 'mate,' so to speak:

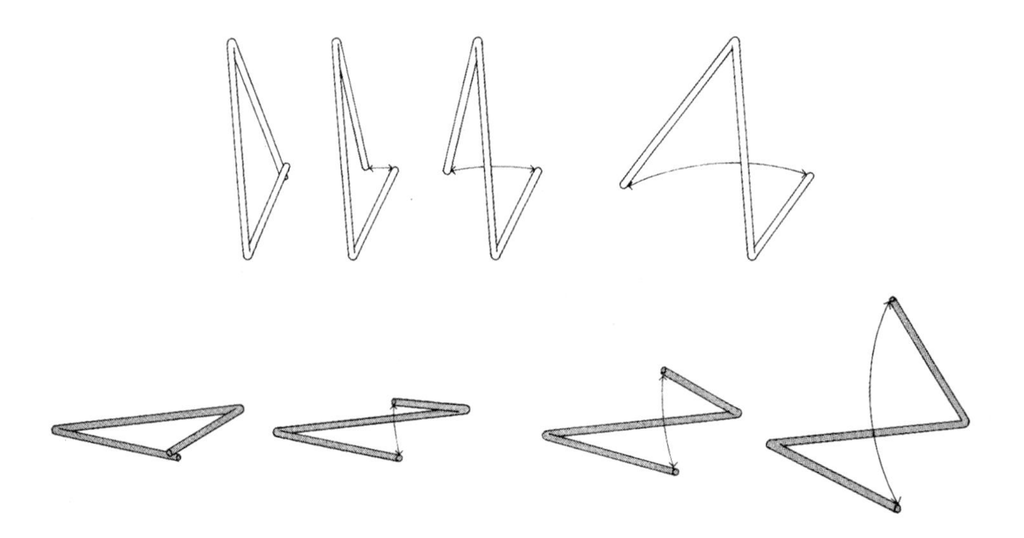

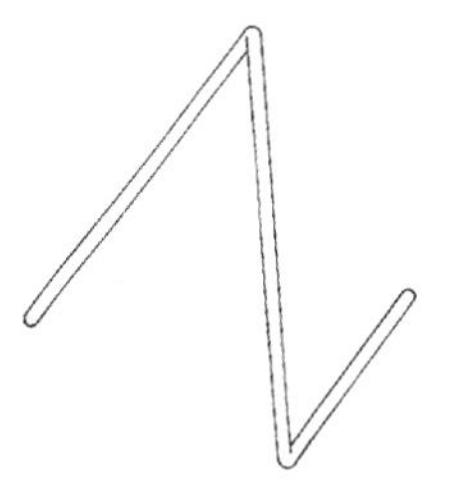

and its opposite event

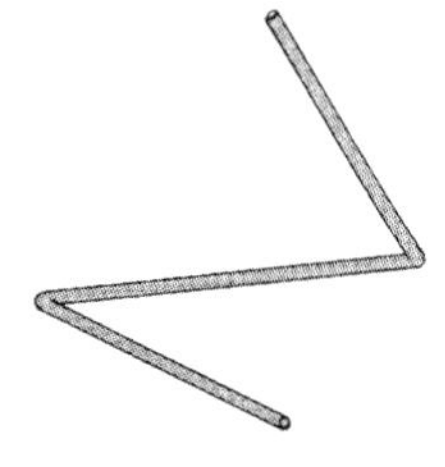

become

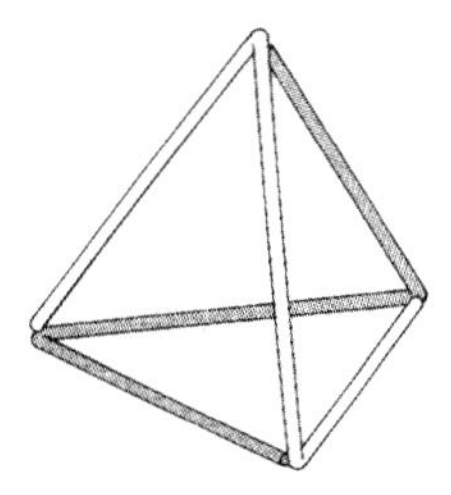

Here something perfectly astonishing has happened. One triangle has fused with another to create four triangles: 1 + 1 = 4. Quite vividly the whole is more than the sum of its parts.

This new creature now 'sticks out' (literally *ek-sists*) in the world – it has an *inside* and an *outside* – and with it come all the ins and outs of the myriad relationships we experience in the world.

How are we to understand this newborn creature, made up solely of triangles and yet *more* than a triangle?

2. Tetrahedron

Thing and No-Thing

What we have discovered is the *tetrahedron* (Greek, *tetra,* four, *hedron,* side, face), the minimum 'anything' with an inside and outside. Nothing simpler than the tetrahedron exists and, as Fuller maintains, everything that exists is tetrahedrally coordinate.

How dare anyone make such a bold assertion?

- Any 'thing' must have an inside and an outside (itself and the rest of the Universe) in order to be a 'thing.'
- The minimum number of events which defines an inside and outside is four.
- A fourfold relationship of events is always a tetrahedron.

Therefore if 'things' are but the relationships between or among energy events, then the tetrahedron is the most economical path Nature can take to create any 'thing.'[2] This should not imply that the tetrahedron is the single 'building block' or only explanation of the world. No single perspective will give you the whole of it, as we have already observed. Yet tetrahedral coordination displays some curiously universal properties which give it precedence over other energetic configurations when we try to account for the shape(s) the world is in.

Tetrahedral Accounting

Strictly speaking, the cube is not a structure; to use Fuller's language, it is not a "self-stabilizing energy event complex."

> *If you try to account in cubes for nature's energy associabilities – as structural systems – you use up to three times as much space as you do if you count space volumes in tetrahedron units.*[3]

One would not expect energy to take the long way around, and it never does. Nature is always most economical in structure.

Let's take a look at what this cube is really made of:

- The diagonals of the faces of a cube are the six edges (vectors) of a tetrahedron

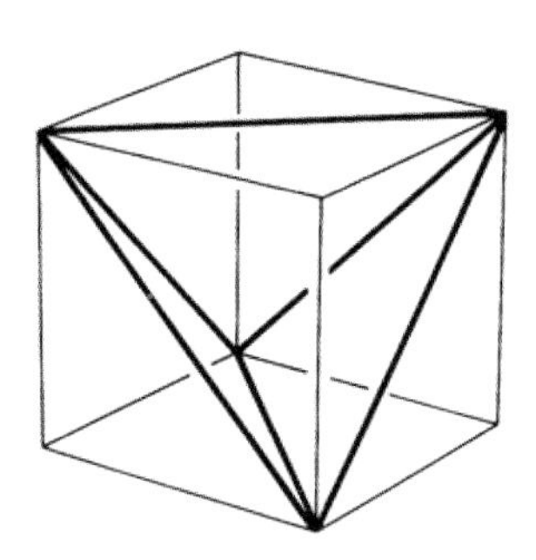

- Volumetrically, a cube is no more than a tetrahedron with four smaller tetrahedra attached to each of its four faces.

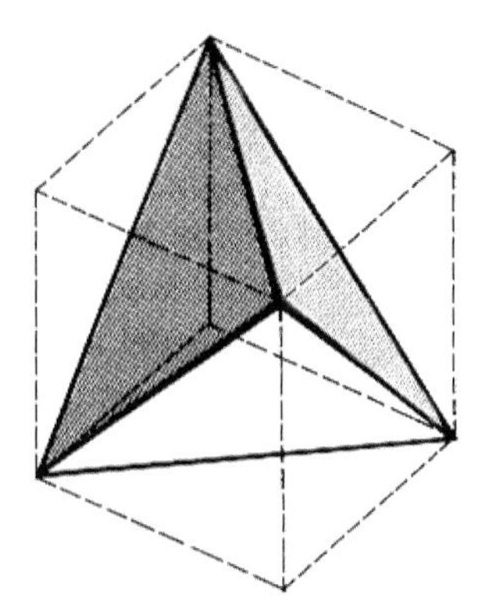

- Structurally, a cube is nothing more nor less than two equal tetrahedra (one positive, one negative) joined at their common centers.

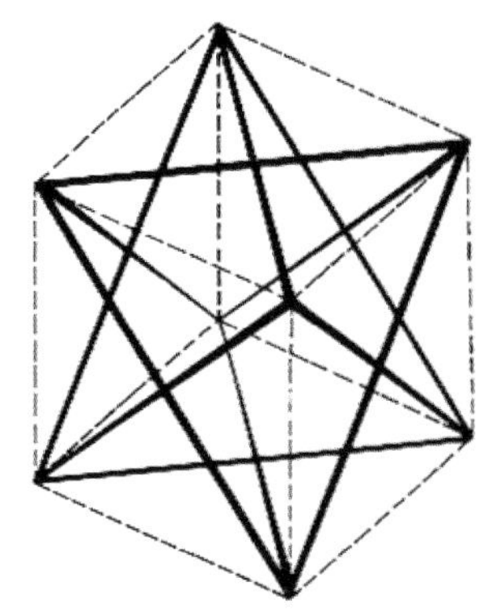

So if you have ever wondered how complex and irrational numbers came to be so complex and irrational, the culprit is the cube. Applied willy-nilly to natural phenomena, the cube invariably distorts whatever it is used to account for. It is an arbitrary, abstract and misleading model of nature's own very definite structural strategies.

Everywhere and Nowhere

Another way to get at all this: The simplest self-stabilizing arrangement of spheres also reveals itself to be (as Bucky liked to say) "our friend the tetrahedron." Here you have in a nutshell the 'omnitriangulated' set of relationships Fuller will draw out into Nature's own coordinate system: "Tetrahedron Discovers Itself and Universe." (¶480.00)

Picture these closest-packed spheres multiplying:

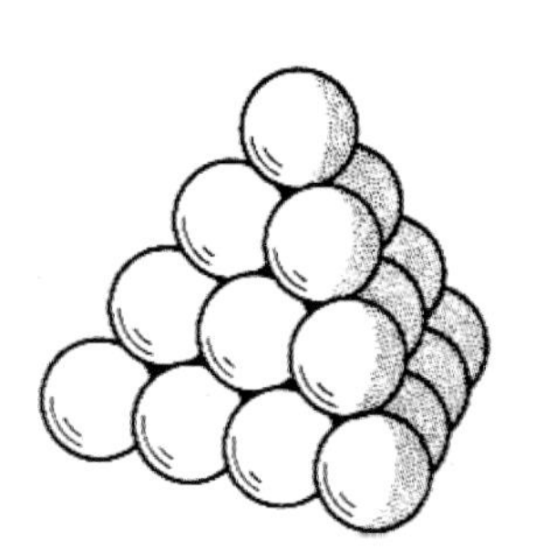

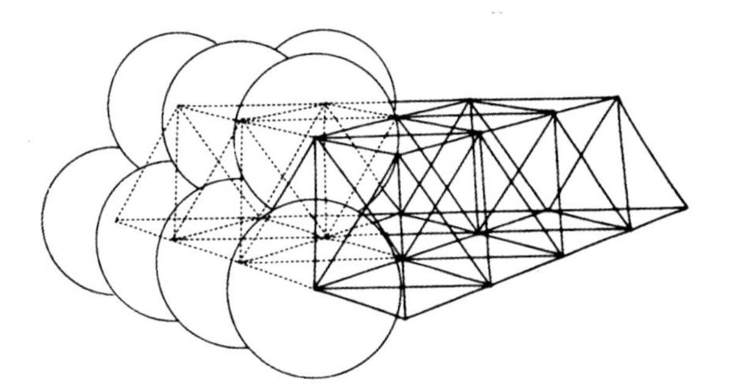

Each sphere in this matrix will eventually be surrounded by twelve other spheres at equal distances, each of those twelve will be at the center of twelve others, and so on.

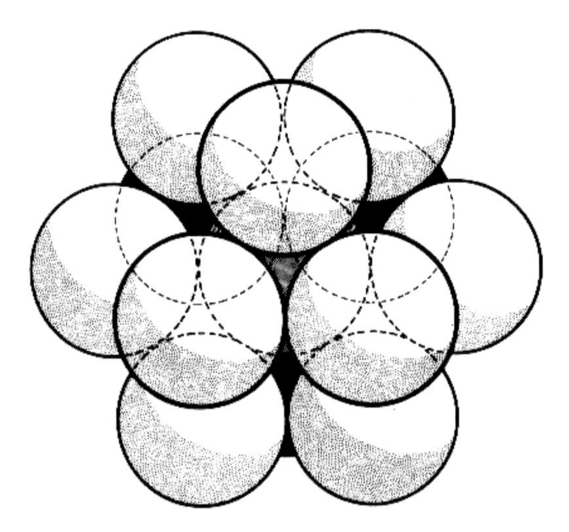

The center is everywhere, and nowhere.

The vectors connecting the center of each sphere to the centers of the spheres immediately surrounding it form an *isotropic vector matrix* (Greek, *iso,* same; *tropos,* turning), so called because not only are all its vectors equal in length, but the angles around any convergence are always identical: 60-degrees. In traditional physics, combined linear and angular momentum can only be described in terms of unresolvable square roots, because the chords and radials of the XYZ coordinate system are the hypotenuses and legs, respectively, of right triangles. In synergetics, however, the isotropic vector matrix provides for rational, whole-number accounting. As Fuller puts it, this matrix is

> *... the omnirational accomodation of both linear and angular acceleration in the same mathematical coordinate system ... [because its] fundamental 60-degree coordination operates either circumferentially or radially. This characteristic is lacking in 90-degree coordination, where the hypotenuse of the 90-degree angles will not be congruent and logically integratable with the radials. (¶423.03)*

The isotropic vector matrix 'gives birth' (Latin, *matrix*, womb; as in *mater*, mother) to all the associative (syntropic) and dissociative (entropic) transformations in Universe. It is implicit in the Law of Conservation of Energy, which holds that energy can neither be created nor destroyed. As such, it may well figure in modern physics' century-long quest for the so-called 'unified field':

> *The synergetics system expresses divergent radiational and convergent gravitational, omnidirectional wavelength and frequency propagation in one operational field. (¶982.52)*

In other words, through this matrix it may well be possible to see Nature *whole*. One should, however, bear in mind Fuller's proviso that he is not 'copying' Nature, but learning to build things the way she does:

> *I did not copy nature's structural patterns. I began to explore structure and develop it in pure mathematical principle, out of which the patterns emerged in pure principle. I then applied them to practical tasks. The reappearance of [such] structures in scientists' findings at various levels of inquiry confirms the mathematical coordinating system employed by nature. (¶203.09)*

Bucky first encountered the isotropic vector matrix in kindergarten (1899) while playing with toothpicks and semi-dried peas. Nearly blind before receiving his first pair of glasses, he *felt* for the structure that would hold itself together. First he made triangles, these became tetrahedra, and by adding more tetrahedra tip-to-tip little Bucky built up a trusswork of tetrahedra, with octahedral spaces popping out between them.[4] Because of its alternating tetrahedra and octahedra he later nicknamed this matrix the 'oc-tet truss.' He early on sought a patent for this structure, but was informed that he could not patent a geometrical structure, only applications made from it. Besides, Alexander Graham Bell had already designed and built a gigantic 'tetra-kite' out of oc-tet modules – and it flew! Still, you can spot this truss in all sorts of engineering projects all around you: in the long necks of the cranes used to build (alas, rectangular) skyscrapers, for instance, or the open tubular metal roofing structures so often used for airports and service stations.

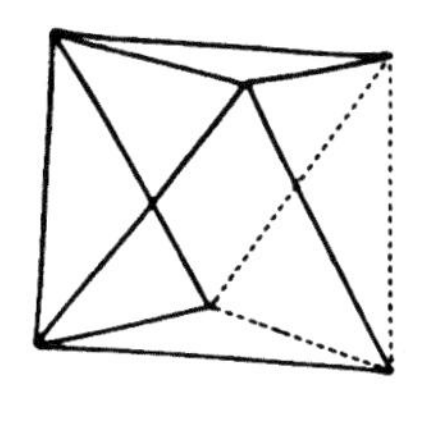

octahedron

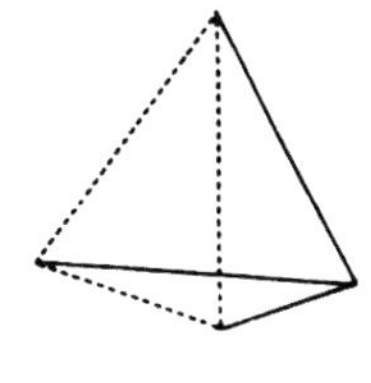

tetrahedron

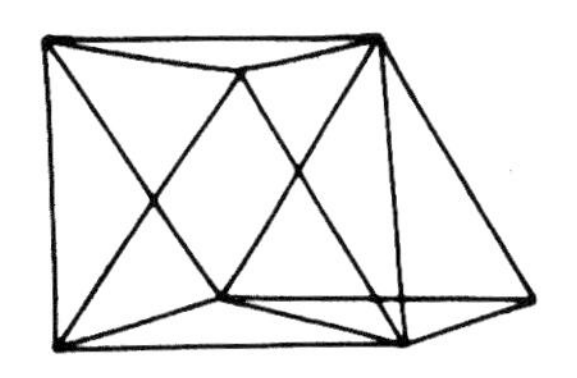

octahedron-tetrahedron truss (oc-tet truss)

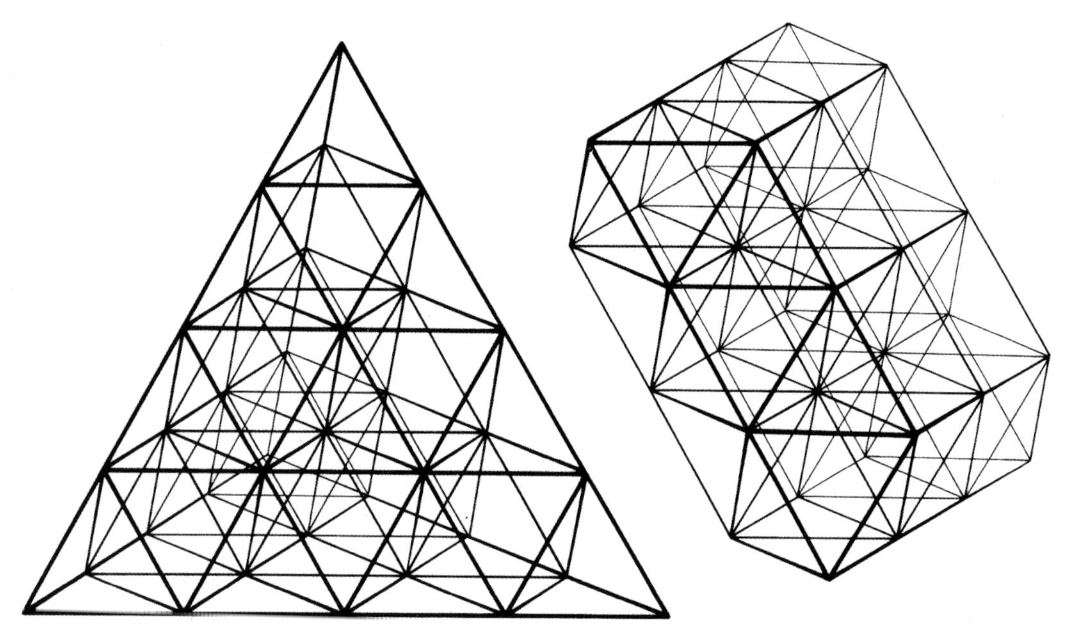

Have you ever tried to visualize the fourth dimension? You've always been frustrated because the classical model of the fourth dimension is an ungainly creature called 'hypercube' – a cube turned inside out in directions perpendicular to its faces. You can draw a two-dimensional representation of what this three-dimensional model of the fourth dimension should look like –

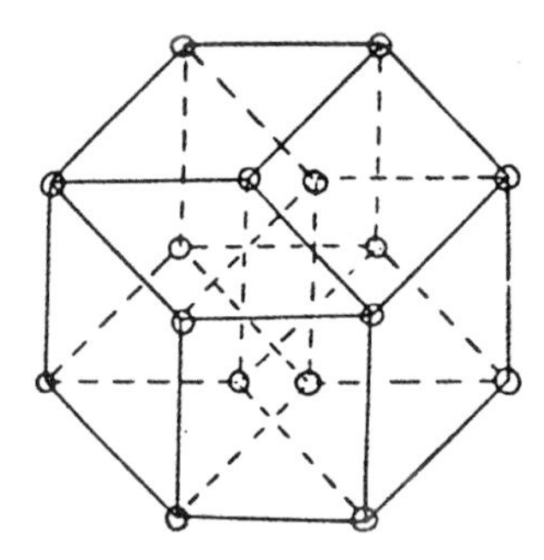

– but it is impossible to actually build a three-dimensional model of the legendary hypercube. Interestingly, when we proceed with the same strategy, but using the tetrahedron instead of the cube, our luck changes. As E. J. Applewhite points out: "What the three axes of the cube do for three dimensions, the four axes of the tetrahedron do for four dimensions."[5]

In this sense, you need not be surprised when you turn to the models covered in the next few pages, or turn your hand to constructing them, and find yourself making 'four-dimensional' artifacts. The four axes of the tetrahedron orient us in four *spatial* dimensions, of course, but it is almost also fair to say that the dimension of *time* has indeed been added – a topic we shall take up in more detail under the heading of 'frequency' a little later. All of the WHOLES to follow in this Appendix are in Fuller's terminology 4D models, but some will take you a little more 'time' than others. The first ones, however, are quite simple and easy to construct.

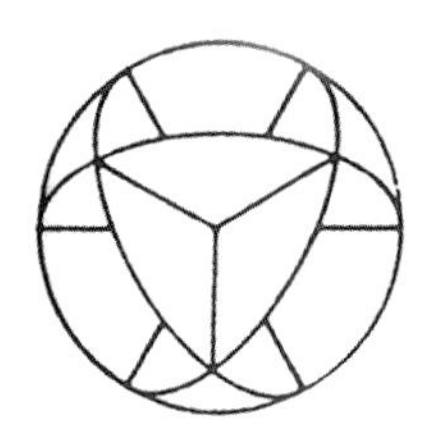

3. Vector Equilibrium

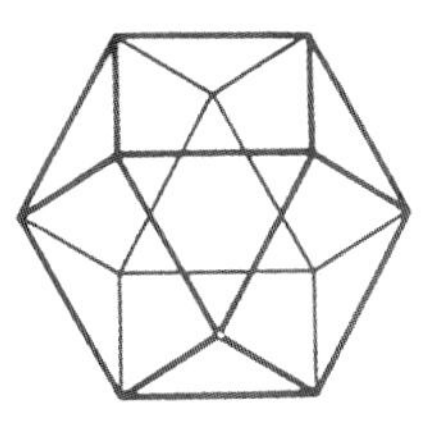

Vector equilibrium is the 'zero' of synergetics, the nucleus. It is so to speak the place of peace at the very center, which is of course not a 'place' at all. Vector 'equilibrium' is the phase through which anything and everything must pass to become itself or anything else, the nexus of any and every energy event.

In Fuller's eyes, the vector equilibrium tells us some very important things about origins, the way things begin: "If it is a starting point, it is a vector equilibrium." (¶440.07) And so we shall start our model-making with this figure, rather than with the spherical tetrahedron and octahedron.

One might assume that the first layer of 12 spheres closest-packed around a central sphere would make yet another (knobbled) sphere. Surprisingly, it does not. It looks instead like this:

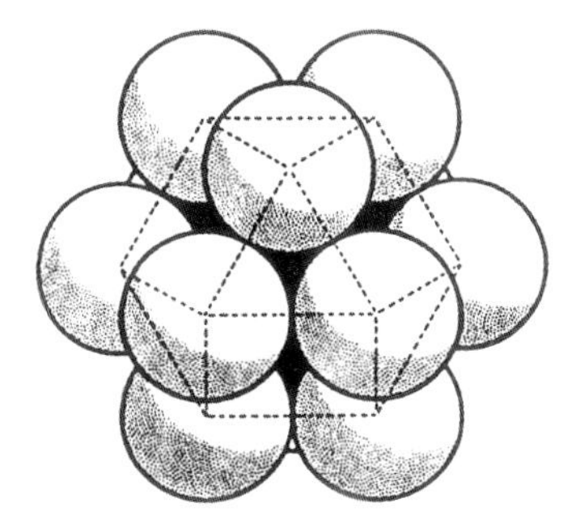

Fuller saw in this figure which emerges from closest-packed spheres not merely another 'solid' phenomenon, but a dynamic complex of energy events. In classical parlance it is known as the 'cube-octahedron,' a name provided by Archimedes, who showed that it can be seen either as a truncated cube:

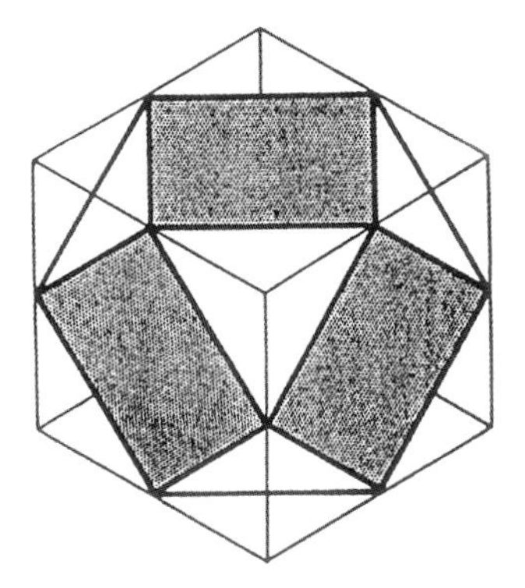

or as a truncated octahedron:

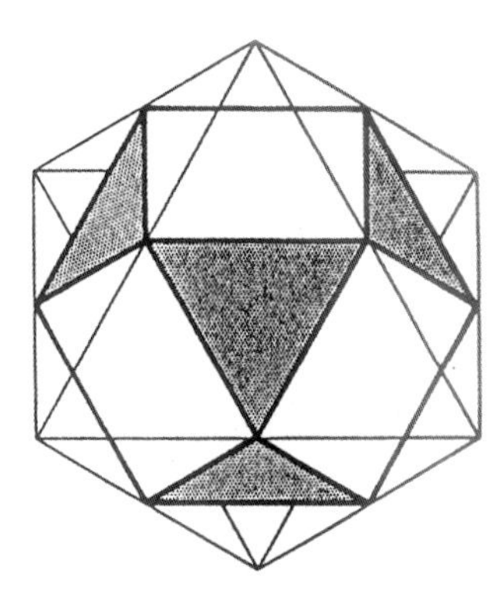

But Fuller's vision cut deeper. He saw that the edge vectors were equal in length to the vectors connecting the center with the surface – i.e., the tensile forces pulling in on the system (the bounding edge vectors) were in dynamic balance with the compressive forces pushing outward from within the system (the radiating vectors). And so he christened this primordial synergy, which plays such a prominent part in atomic structural strategies that it has become the conventional international symbol for atomic power, the *vector equilibrium*.

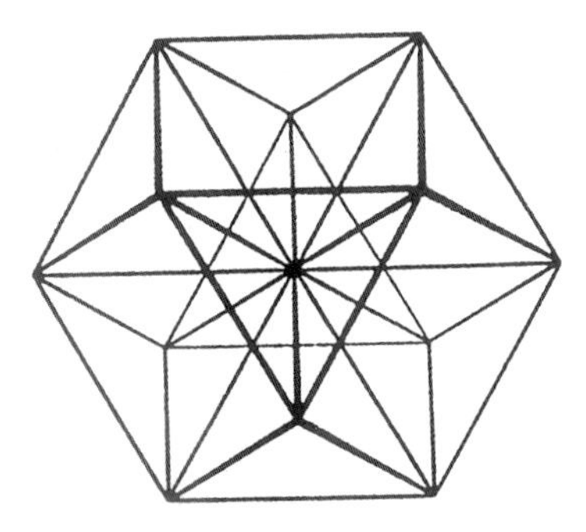

As the circumferentially united and finite great-circle chord vectors of the vector equilibrium cohere the radial vectors, so also does the metaphysical cohere the physical. (¶440.08)

The Vector Equilibrium is the simplest of the WHOLES to construct, but it still surprises everyone who actually assembles one. Unlike some of Fuller's other, better known projects (oc-tet trusses, geodesic spheres, tensegrity figures), there has never been any dispute as to who 'really' invented these foldable geometries. In her excellent redaction of Bucky's geometry, *A Fuller Explanation,* Amy Edmonson marvels at how in the world Fuller ever came up with such a method for making spheres out of folded 'bow-ties':

Fuller made a remarkable discovery about great-circle patterns that is responsible for their great significance in his mathematics. This discovery involves an intricate relationship between central and surface angles and could so easily be missed that one cannot help but reflecting on the intuition that led Fuller to such an insight.[6]

It is a indeed a signal case of Bucky's intuitive method operating to disclose fundamental structural insights – perhaps first achieved with the vector equilibrium, since its central and surface angles are exactly the same. Another clue as to how he arrived at these exquisitely beautiful figures may well be that WHOLES model electromagnetic wave phenomena: spherical wave growth out from centers. He himself explicated what is going on in such figures as simply an illustration of this principle:

> *It is characteristic of electromagnetic wave phenomena that a wave must return upon itself, completing a 360-degree circuit. The great circle disks folded or flat provide unitary wave-cycle circumferential circuits. Therefore, folded or not, they act like waves coming back upon themselves in a perfect wave control. We find their precessional cyclic self-interferences producing angular resultants that shunt themselves into little local 60-degree bow-tie 'holding patterns.' The entire behavior is characteristic of generalized wave phenomena.* (¶455.21)

In a deadpan way, as if it were the most ordinary phenomenon in the world, this passage outlines the novelty of what is happening in these models. When you unfold the first of these WHOLES, the Vector Equilibrium on the next page, you will begin with a set of circles ... and finish with what appears to be the same set of circles, marvelously transformed into a structurally sound (yet volumeless) sphere. Amazed? So are most people. Amy Edmonson again: "Looking at the finished model ... the procedure is reminiscent of the magic trick in which a handkerchief is cut into tiny pieces and thrown randomly into a hat, only to reappear intact."[7]

As you unfold this Vector Equilibrium and the WHOLES to follow, bear in mind the fundamental synergetics axiom that any WHOLE is always *more* than the sum of its parts. That extra 'ingredient' is the metaphysical integrity of pure principle, the synergy of every part working with every other, which is really all that holds the WHOLE together. So follow the instructions, but not blindly. Allow the emerging pattern of the WHOLE to guide you, and you won't go wrong.

Snow crystals: Vector Equilibrum

Unfolding the Vector Equilibrum

4 disks
12 pins (see p. 183)

WHOLES are made from scored disks which resemble bow-ties when folded and pinned. The "bow-ties" are then pinned together to form a sphere.

To form the bow-ties:

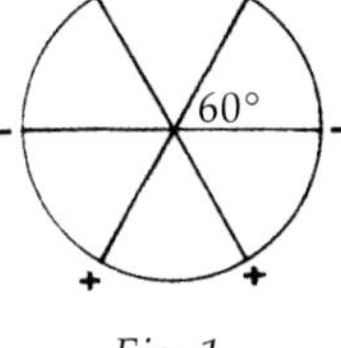

Fig. 1

1) Fold each disk in half along each diameter, as shown in Fig. 1. (Make sharp creases).
   - = fold away from you
   + = fold toward you

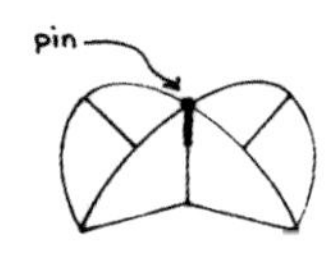

Fig. 2

2) Pin each bow-tie as shown in Fig. 2.

To form the sphere:

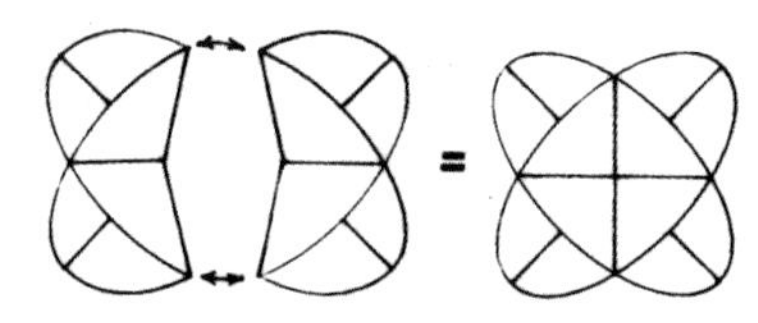

Fig. 3

3) Pin the first two bow-ties together as shown in Fig. 3.

4) Continue this pattern to pin the remaining bow-ties. Fig 4 shows the complete sphere with one bow-tie in relief.

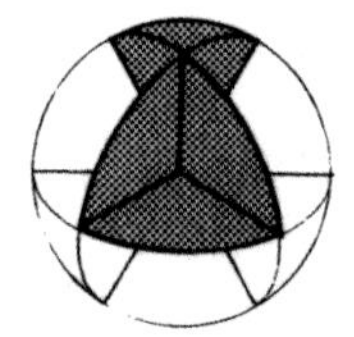

Fig. 4

Note: You can expect the bow-ties to slip a little as you pin them, but the figure will adjust itself with the last pin or two. Pinch firmly on the bow-ties to place the last few pins.

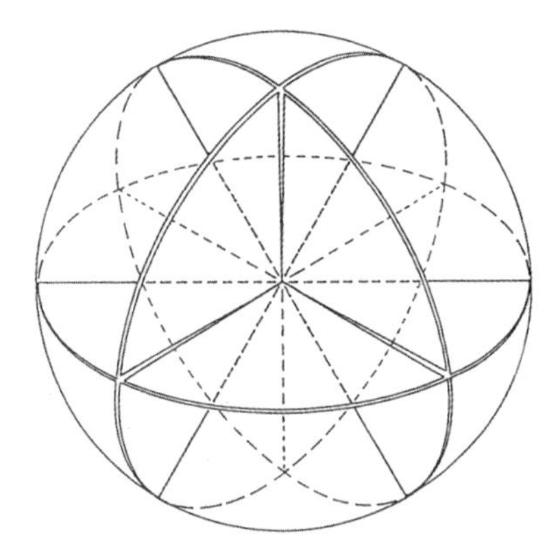

Spherical Vector Equilibrum. © Tom Parker

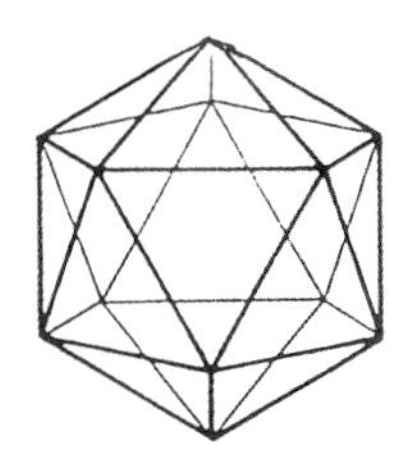

4. Icosa

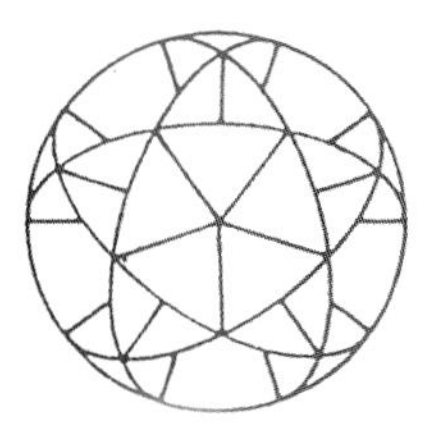

Nature unfolds in certain patterns. The icosahedron (Greek, *icosa*, twenty (triangles)) reveals her integral strategy for most living organisms. Whether it be a leaf popping out from a stem, a nautilus spinning its spiral home in the sea, or the five digits of your own hand, there lies the icosa, shell and shelter, the domicile of life. The icosahedron reveals some of the intimate secrets of organic growth, the way living things unfold.

The five-pointed star of the icosa holds the key to the sacred ratio of Ø (*phi*), known since antiquity as the Golden Section, or Divine Proportion:

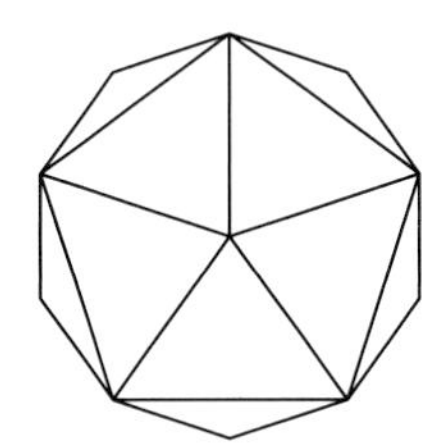

(a) icosahedron

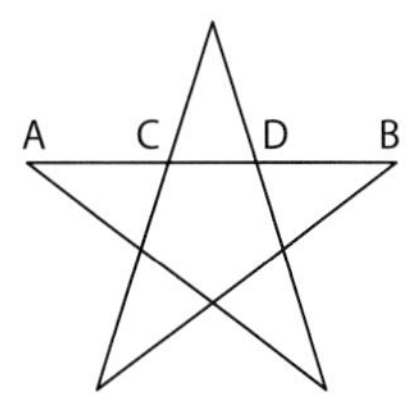

(b) CD is to AD as AD is to AB (the smaller is to the larger as the larger is to the whole) AD/CD = Ø AB/AD = Ø

The icosahedral ratio Ø is also disclosed by the *Fibonacci Series,* where the sum of any two consecutive numbers in the series equals the number following:

0,1,1,2,3,5,8,13,21,34 …

and where any number divided by the preceding number yields the ratio Ø:

13/8 = 1.625 21/13 = 1.6154 and so on …

Consider plants: *phyllotaxis* (the arrangement of leaves on stems) is neatly proportioned to Ø:

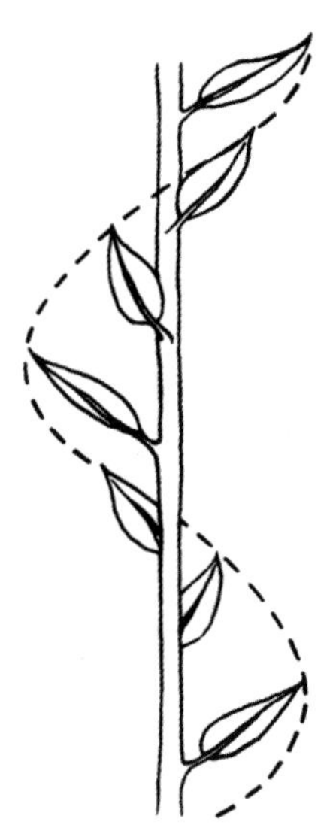

(a) Phyllotaxis. If p is the number of turns of the helix and q is the number of stems passed, then p/q expresses leaf distribution. Both the numerator and the denominator of this fraction are always members of the Fibonacci series

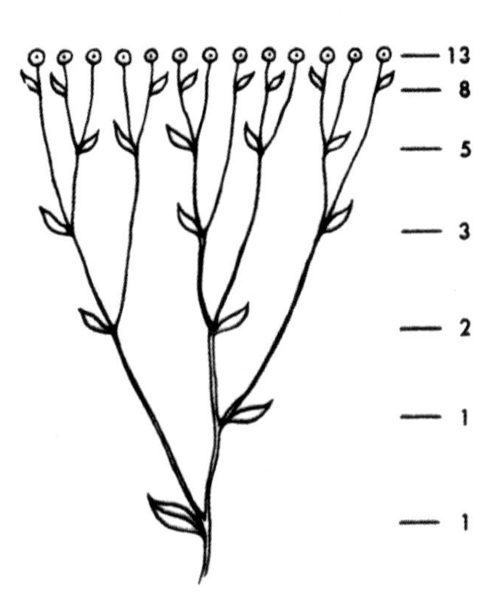

(b) Sneezewort (Achillea ptarmica)

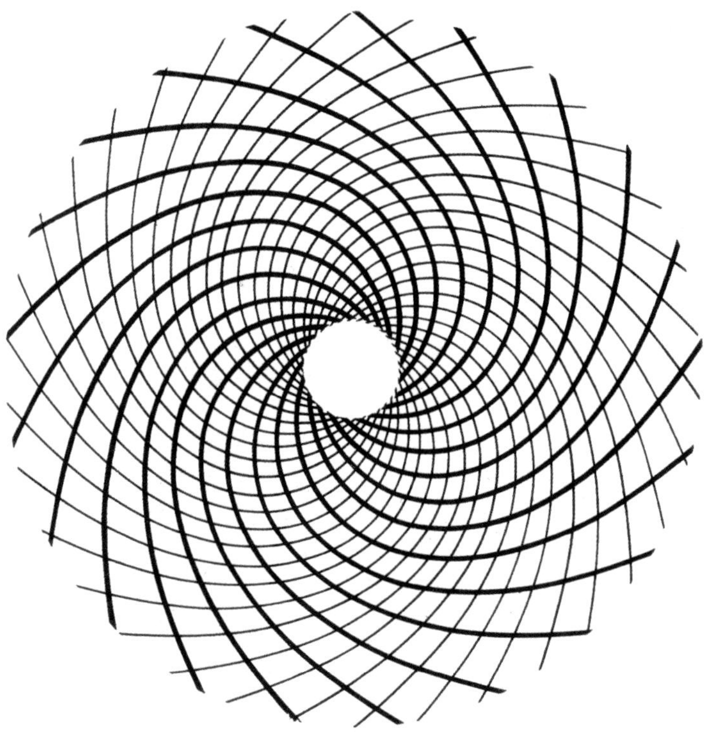

(c) Sunflower. The ratio of clockwise to counter-clockwise spirals is always Ø. These interlocking spirals appear wherever we find leaves, kernels, buds or other floral patterns. See, for example, the kernels on a pine cone or the buds on a pineapple. (H. E. Huntley, The Divine Proportion, *New York, 1970.)*

Ø is also the proportion according to which a snail's shell invariably curls:

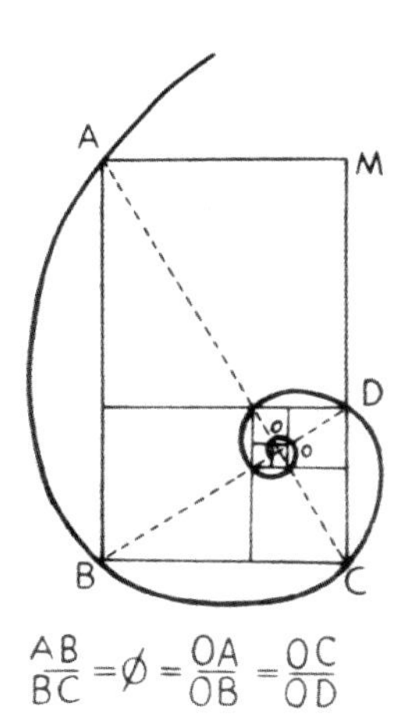

"The architect of the future will build imitating Nature, for it is the most rational, long-lasting and economical of methods" – Antoni Gaudi, cited in the nave of his Sagrada Familia in Barcelona. Photo © Pere Vivas

1.618 is a very accurate numerical approximation of the Ø ratio. Even people are not exempt from this inbuilt proportioning, as Leonardo and others have outlined:

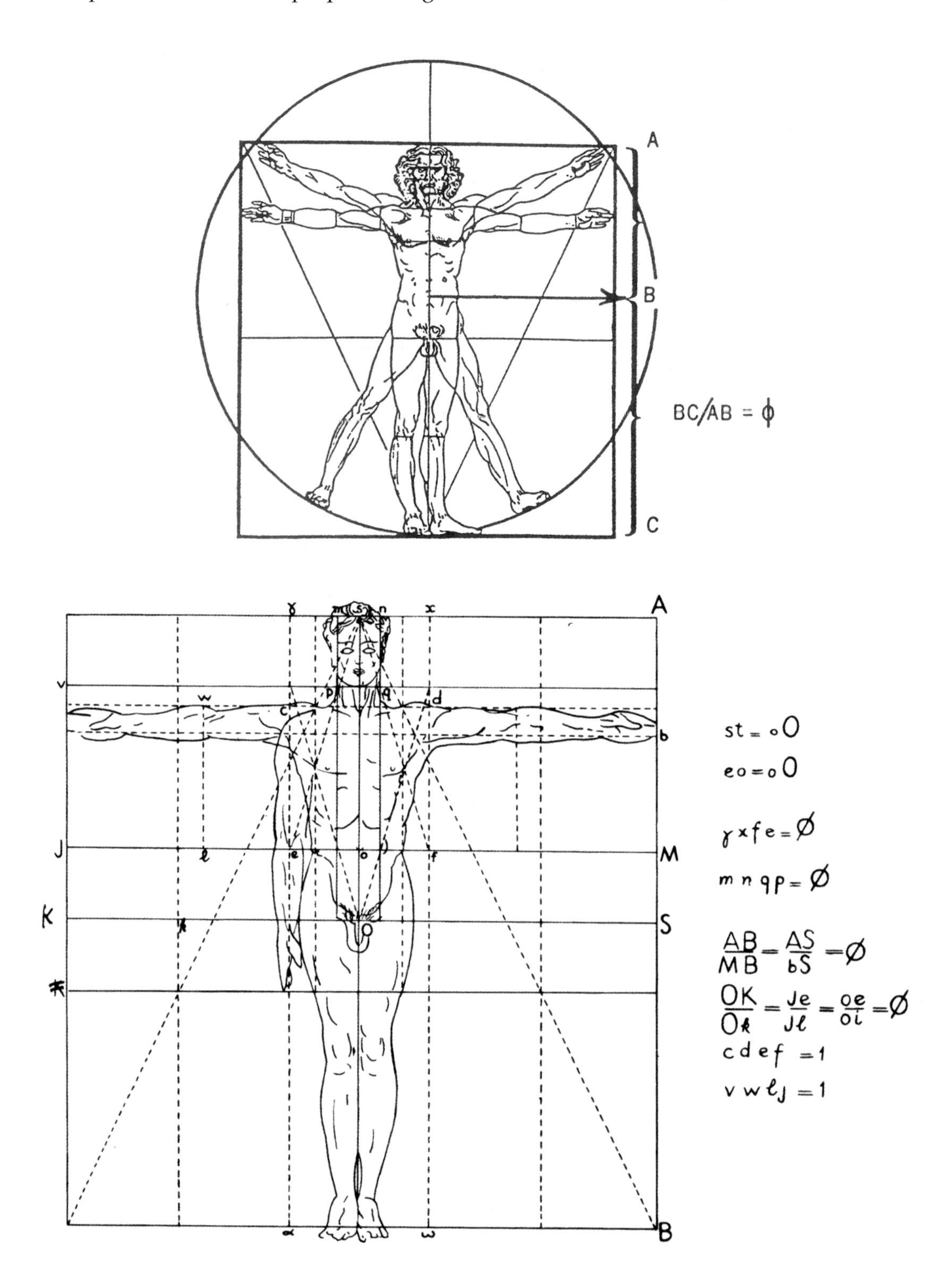

It is not therefore so surprising that the cathedrals of the High Middle Ages in Europe were constructed from polyhedral icosa coordinates, largely incorporating Ø and other sacred ratios.

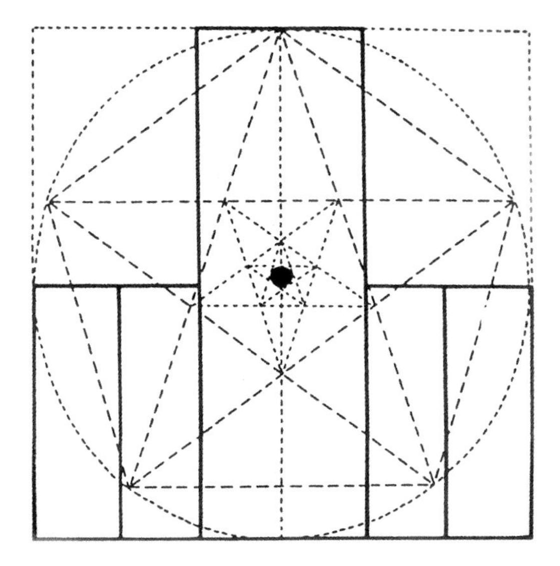

(a) *A diagram showing a vertical section of the nave, from the Gothic cathedral of Cologne.*

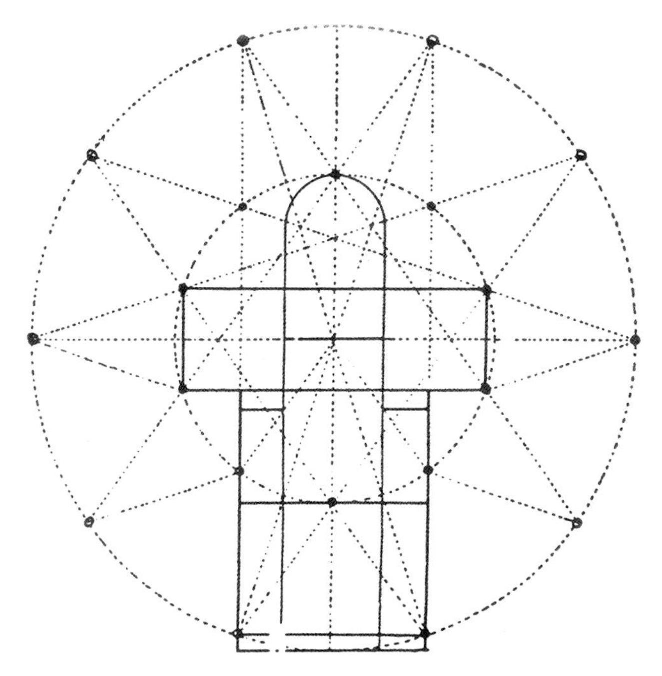

(b) *The Gothic Standard Plan uses a ten-way division of the circle in equal 36-degree increments, as do the disks you will make for constructing the spherical icosa WHOLE.*

By the same token, it is perhaps more than coincidence that Fuller's famous geodesic domes most commonly employ the same icosahedral strategy. Fuller would say that whenever "maximum volume enclosure per unit of invested energy" is the principle function to be served, then Nature uses the icosahedron. This is why the pneumatic and hydraulic structuring of Nature (e.g., trees, flowers, radiolara, protein shells, etc.) employ spherical icosahedral geometry.

An icosahedral geodesic dome places about 1/50th the weight on its foundation as a cubical building of comparable space-enclosing capacity. And geodesic domes enclose space without using any internal supports. Their strength resides in the *integrity of the pattern* cohering the shell, not in the number of bricks, bolts, crossbeams, and so on. As an important side-effect, the geodesic dome is the only manmade object whose structural strength increases as it gets larger. Ask yourself, what was left standing after the 1906 San Francisco earthquake, or the atomic bombing of Hiroshima? Only the domes. And Bucky's geodesic structures are stronger still.

The geodesic dome is a triumph of pattern integrity over brute force: not exactly mind over matter, but mind and matter working together.

Kaiser 145-foot geodesic dome, manufactured in Oakland, California, and erected in 22 hours in Honolulu, Hawaii. At the 22nd hour, the Hawaiian Symphony Orchestra entered; the concert was completed withing 24 hours of the Honolulu landing of the dome's components. The orchestra conductor pronounced the acoustics as "the best is his experience". (Estate of R. B. Fuller.)

Corroborations

- Photographs of common antibody cells taken by Elias Lazarides and Paul Revel in 1979 revealed their 'geodome' structure, confirming Fuller's intuition that Nature employed icosahedral building strategies in the molecular realm.

- Dr. Aaron Klug of Birkbeck College at London University, winner of the 1982 Nobel Prize in Chemistry, and Dr. Ronald Casper of Boston Children's Hospital have found the protein shells of certain viruses to be constructed on the same patterns as geodesic domes.
- Dr. John E. Hauser of the University of California at San Francisco photographed the inside of the synaptic clefts of muscle cells (where the nerve cells 'communicate' with the muscles), and they turned out to look just like Bucky's domes.
- In 1999, Harvard cytologist Donald Ingber proposed that the key to the structure of cytoplasm – the squishy substance that holds your every cell together – lies in tensegrity structures of the sort Kenneth Snelson developed from Fuller's geometry.

Beyond these corroborations from the sciences, the synergetic strategies of geodesics may yet prove to be one of the great foundational discoveries for human architecture, like the segmented arch or the ogive vault (itself actually a hemispherical oc-tet module in stone). Of course, an entire neighborhood of identical domes might not be much of an improvement on the "little boxes made of ticky-tacky" which "all look just the same" in Malvina Reynolds' famous lament over the conformity of Daly City, California. All art – music, design, poetry, and architecture alike – exists in the interplay between fixed patterns and flexible variations. In the final analysis, Bucky Fuller was probably more of a visionary structural engineer than an architect; his primary concern was the efficiency promised by geodesic domes. But a contemporary architect of vision like Norman Foster, who seems to have absorbed Bucky's lessons about triangulated structures as well as his ecological priorities, can go further. Departing from the purely spherical structures Bucky favored, he has found all sorts of intriguing new ways to utilize Nature's hexagonal coordinates – in his sleek and shimmering 'Gherkin' in London, for instance, or the bold new Hearst Tower in New York City.

The interconnecting mushroom 'Biomes' of Britain's Eden Project in Cornwall illustrate what can be built from Fuller's 'hex-pent' geodesic forms with today's lightweight, resilient materials and the ultra-precise machining permitted by computer-assisted design programs. As you unfold the next WHOLE, the Spherical Icosa from which all these beautiful domes are derived, you may begin to appreciate why they look as well as they work. This is not just elegant structure, it is the very structure of elegance.

Unfolding the Icosa

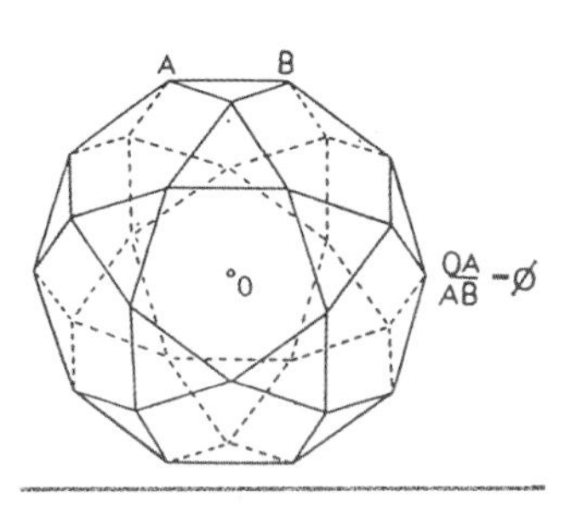

6 disks
30 pins (see p. 182)

To form the bow-ties:

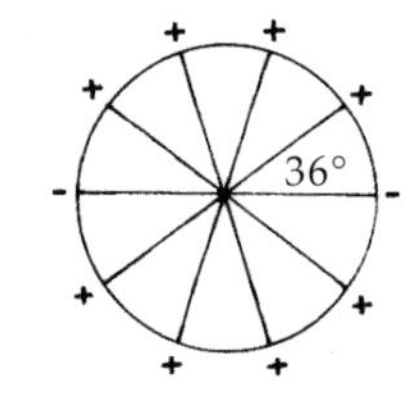

Fig. 1

1) Fold each disk in half along each diameter, as shown in Fig. 1. (Make sharp creases).
   - = fold away from you
   + = fold toward you

2) Pin each bow-tie as shown in Fig. 2.

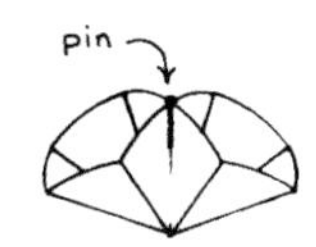

Fig. 2

To form the sphere:

3) Pin the first two bow-ties together as shown in Fig. 3

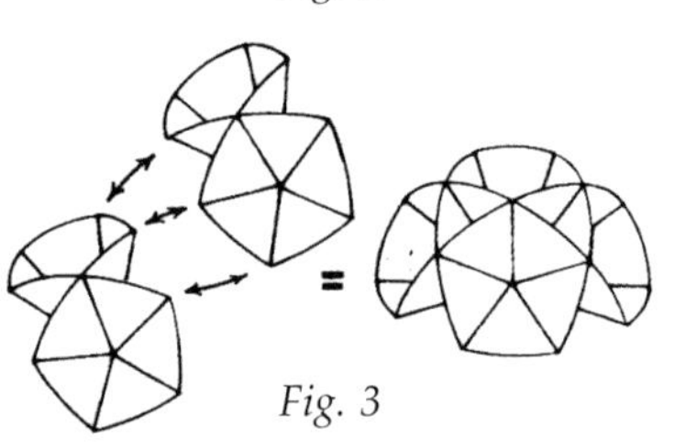

Fig. 3

4) Continue this pattern to pin the remaining bow-ties. Fig 4 shows the complete sphere with one bow-tie in relief.

Note: You can expect the bow-ties to slip a little as you pin them, but the figure will adjust itself with the last pin or two. Pinch firmly on the bow-ties to place the last few pins.

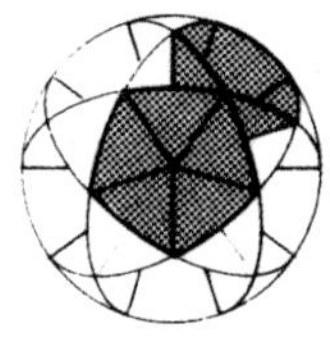

Fig. 4

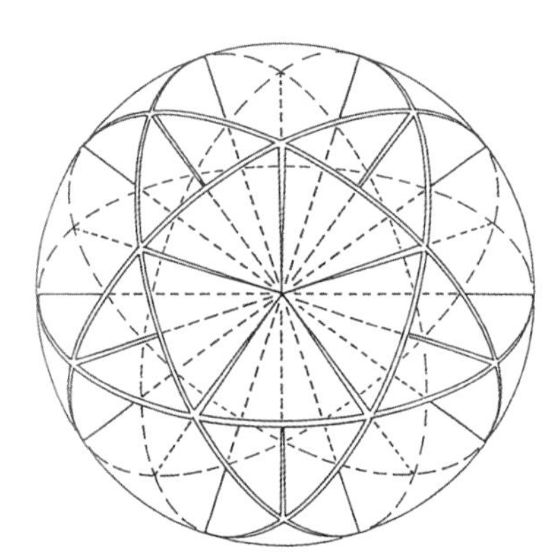

Spherical Icosadodecahedron. © Tom Parker

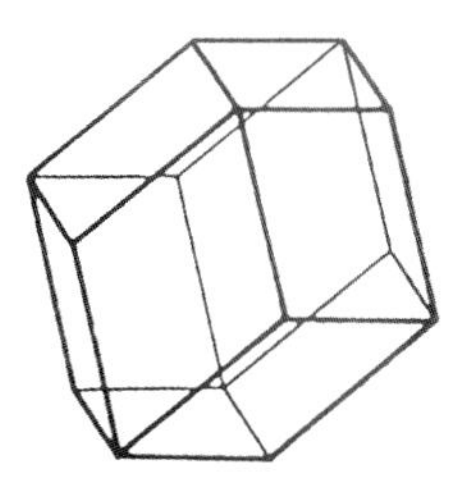

5. Rhombic Dodecahedron

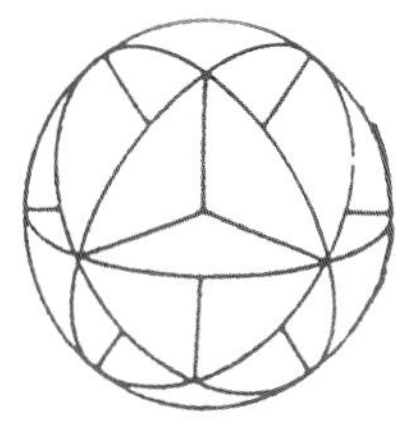

The ancient geomancer's riddle – the 'squaring of the circle' – cannot be resolved in two dimensions. (Rupturing the flat planes of the Euclidean mentality may very well have been the point of posing the conundrum.[8]) But in the spherical rhombic dodecahedron (Greek, *dodeca*, twelve (rhombs)) we witness the *sphering of the cube:* the six great circles diagonally bisect the cube's six faces.

Once you begin assembling this third WHOLE, you will see just how beautifully it transforms the cube into a sphere:

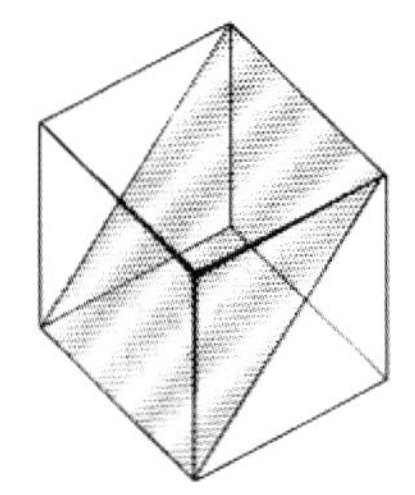

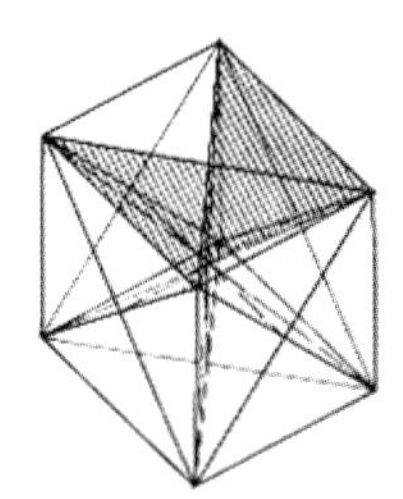

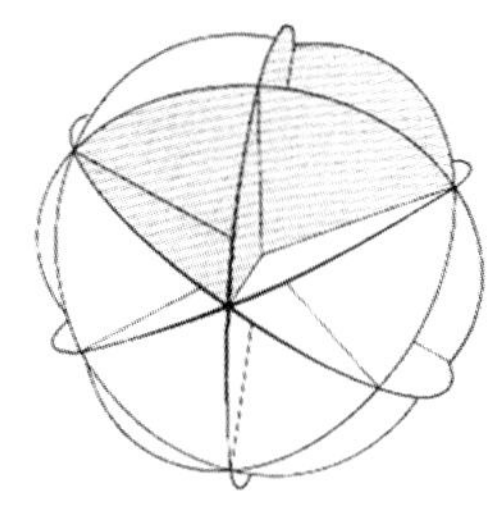

In this model, sphere and cube are reconciled; the extremes meet, and embrace. The rhombic dodecaheron is an *allspace filler.* Fuller notes that "of all the polyhedra nothing falls so easily into a closest packed group of its own kind as does the rhombic dodecahedron." (¶955.52)

The rhombic dodecahedron is also the pattern which best accomodates the transmission and reception of electromagnetic waves. The first radio tuning crystal was presumably a rhombic dodecaheron, that is, as symmetrical a bit of quartz crystal as could be found for somebody's foxhole radio. This figure, then, may tell us something about *attunement*, the way things – and/or people – relate to one another most closely.

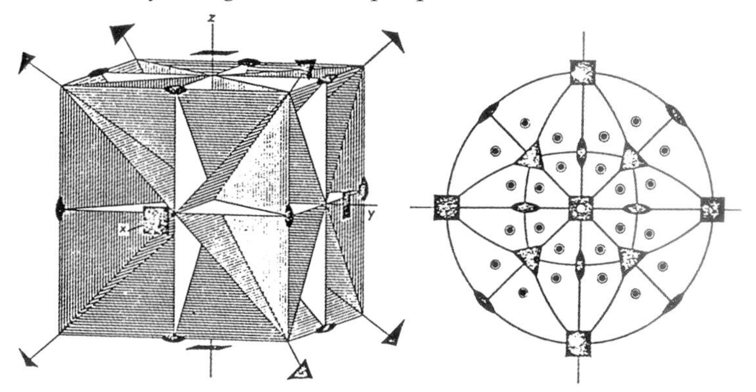

Class m3m; *the elements of symmetry, crystallographic axes and a stereogram of the general form.*
Fig. 303, F. C. Phillips, Introduction to Crystallography, *New York (Wiley) 1971, p.163.*

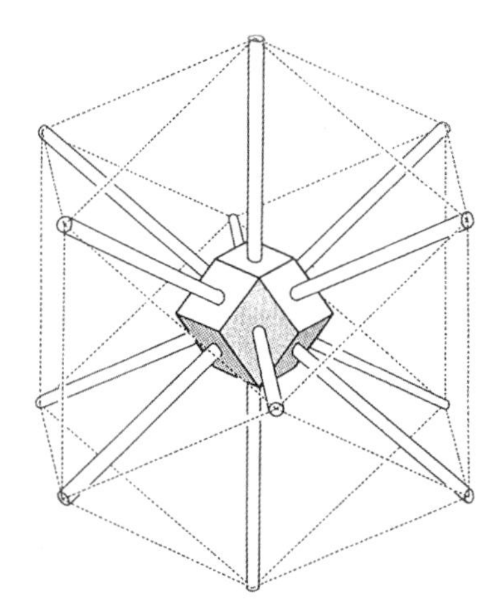

Crystallographers are familiar with the rhombic dodecahedron as a domain of reference with which to account for the growth and structure of natural crystals.

In Fuller's synergetic geometry, the rhombic dodecahedron can be identified as the hub of the vector crossings within the isotropic vector matrix.

Fuller describes this 'centrality' of the rhombic dodecahedron for Nature's coordinate system in an engaging way which also recapitulates much of our presentation of these figures so far:

> *Instead of initiating universal mensuration with assumedly straight-lined, square-based cubes firmly packed together on a world plane, we should initiate with operationally verified reality; for instance, the first geometrical forms known to humans, the hemispherical breasts of mother against which the small human spheroidal observatory is nestled. The synergetic initiation of mensuration must start with a sphere directly representing the inherent omnidirectionality of observed experience. Thus we also start synergetically with wholes instead of parts …*
>
> *We find that the sphere becomes operationally omnicontiguously embraced by other spheres of the same diameter, and that ever more sphere layers may symmetrically surround each layer by everywhere closest packing of spheres, which altogether always and only produces the isotropic vector matrix.*
>
> *This demonstrates not only the uniformly diametered domains of closest packed spheres, but also that the domains' vertexially identified points of the system are the centers of closest packed spheres, and that the universal symmetric domain of each of the points and spheres is always and only the rhombic dodecahedron. (¶981.19)*

To summarize the intimate inter-relationship of these 'duals,' as the rhombic dodecahedron and vector equilibrium are sometimes known, Fuller has recourse to metaphor:

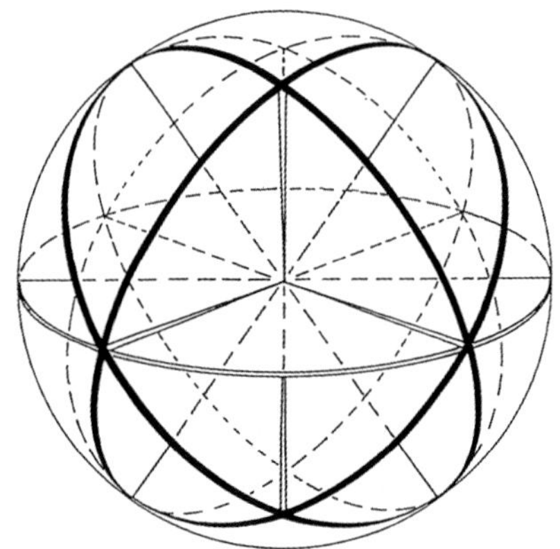

> *The rhombic dodecahedron is symmetrically at the heart of the vector equilibrium. The vector equilibrium is the ever-regenerative, palpitatable heart of all the omniresonant hearts of Universe. (¶984.00)*

You need only hold your model up at a certain angle to see the interlocking heart shapes, which Fuller's metaphor turns into a fitting gnomon of his WHOLE synergetic ('co-operative') universe.

Unfolding the Rhombic Dodecahedron

6 disks
30 pins (see p. 182)

To form the bow-ties:

1) Fold each disk in half along each diameter, as shown in Fig. 1. (Make sharp creases).

  - = fold away from you
  + = fold toward you

2) Pin each bow-tie as shown in Fig. 2.

To form the sphere:

3) Pin the first two bow-ties together as shown in Fig. 3

4) Continue this pattern to pin the remaining bow-ties. Fig 4 shows the complete sphere with one bow-tie in relief.

Note: You can expect the bow-ties to slip a little as you pin them, but the figure will adjust itself with the last pin or two. Pinch firmly on the bow-ties to place the last few pins.

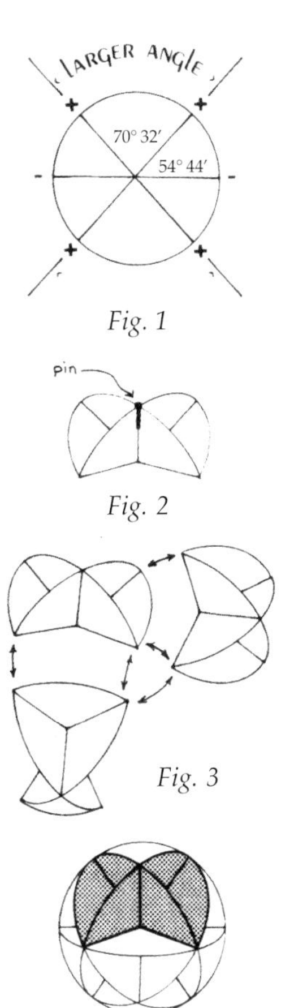

Fig. 1

Fig. 2

Fig. 3

Fig. 4

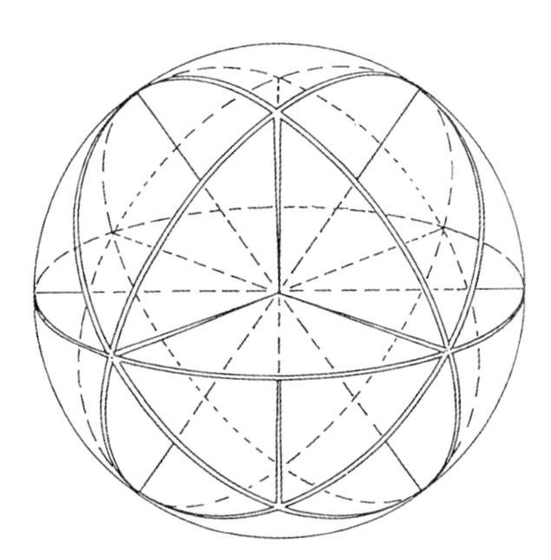

Spherical Rhombic Dodecahedron. © Tom Parker

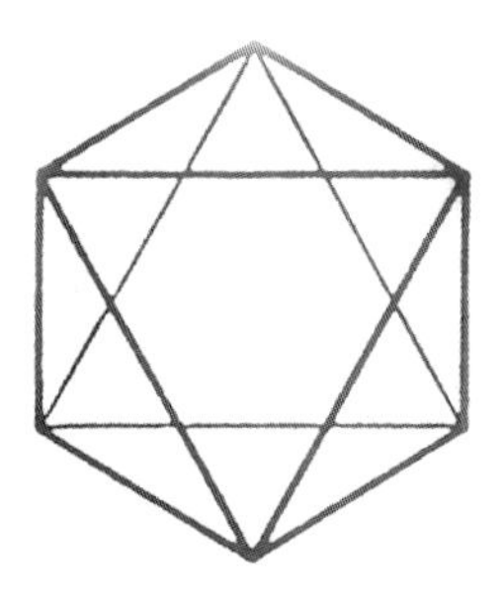

6. Spherical Tet and Octa

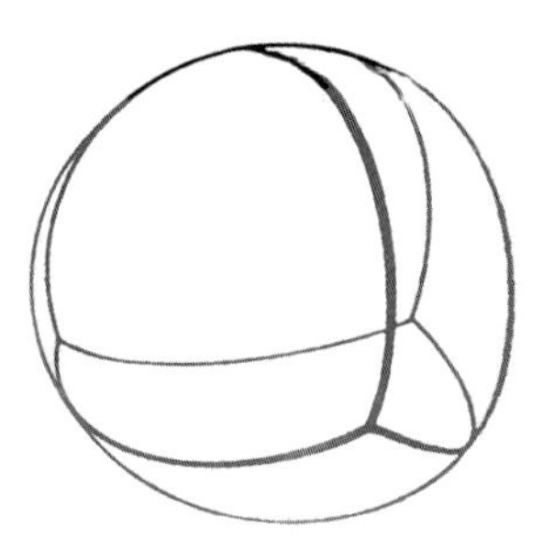

There is one foldable WHOLE we've skipped over, the spherical octahedron (Greek, *octa*, eight). Octahedra appear in the isotropic vector matrix as the voids between tetrahedra. Fuller has covered the construction of the spherical octa in great, not to say overwhelming, detail in *Synergetics* ¶835.00 and ¶836.00.[9] There is no need to belabor its relatively simple construction here. But the reader may also be wondering why we have encountered no spherical *tetrahedron* in our set of foldable figures.

The spherical tetrahedron is not a member of the family of (seven and only seven) foldable geodesic polyhedra, which Fuller was accustomed to calling 'great circle models.' The spherical tet's 109-degrees, 28-minutes of spherical angle do not resolve into equal 360-degree-totalling increments. It is simple enough to make a rough spherical tetrahedron, out of clay for example, but every such model is bound to be unique. If you were to use a mold, since the spherical tet has no continuous diameter, every time you made one you would have to break the mold. Of course there is a ready-made spherical tetrahedron we have all observed without noticing it. The distribution of land masses on our planet Earth forms a rough spherical tet, although only Bucky's Dymaxion Projection shows it in proper proportion.

It can be said, moreover, that the spherical tetrahedron is the *arché* of all the WHOLES presented here, the most primordial pattern, the granddaddy of them all. Look sharply and you'll catch sight of it almost everywhere, but always in mid-transformation, always in the process of turning itself into some other, more highly articulated, pattern.

In his *Cosmic Fishing*, E. J. Applewhite describes their collaborative book *Synergetics* in its entirety as "a business of stark homage to the tetrahedron." Fuller retorts (on the same page!): "I am not in homage to anything, certainly not the tetrahedron as an object, merely as the minimum structural system in Universe."[10]

Returning to the octahedron, which *can* be 'unfolded' as a WHOLE, we find that its 90-degree coordination introduces an element of redundancy (doubled-up edges) not present in the previous figures. Fuller explains:

> *The octahedron always exhibits the quality of doubleness. You might think you could do it with one set of three great circles, but it takes* two *sets of three great circles to fold the octahedron. (¶937.21)*

Crystals are face-bonded molecules. Thus, to paraphrase Fuller, because it is double-bonded and its vectors are doubled, the octahedron may be considered the optimum representation of crystalline structure: the original diamond in the rough. And indeed, the octahedron turns out to be the most common crystal formation in nature.

Here are the simple specifications for the spherical octa. You should have no trouble folding six paper circles in half twice to obtain 90-degree quadrants.

Unfolding the Spherical Octahedron

6 disks
10 pins

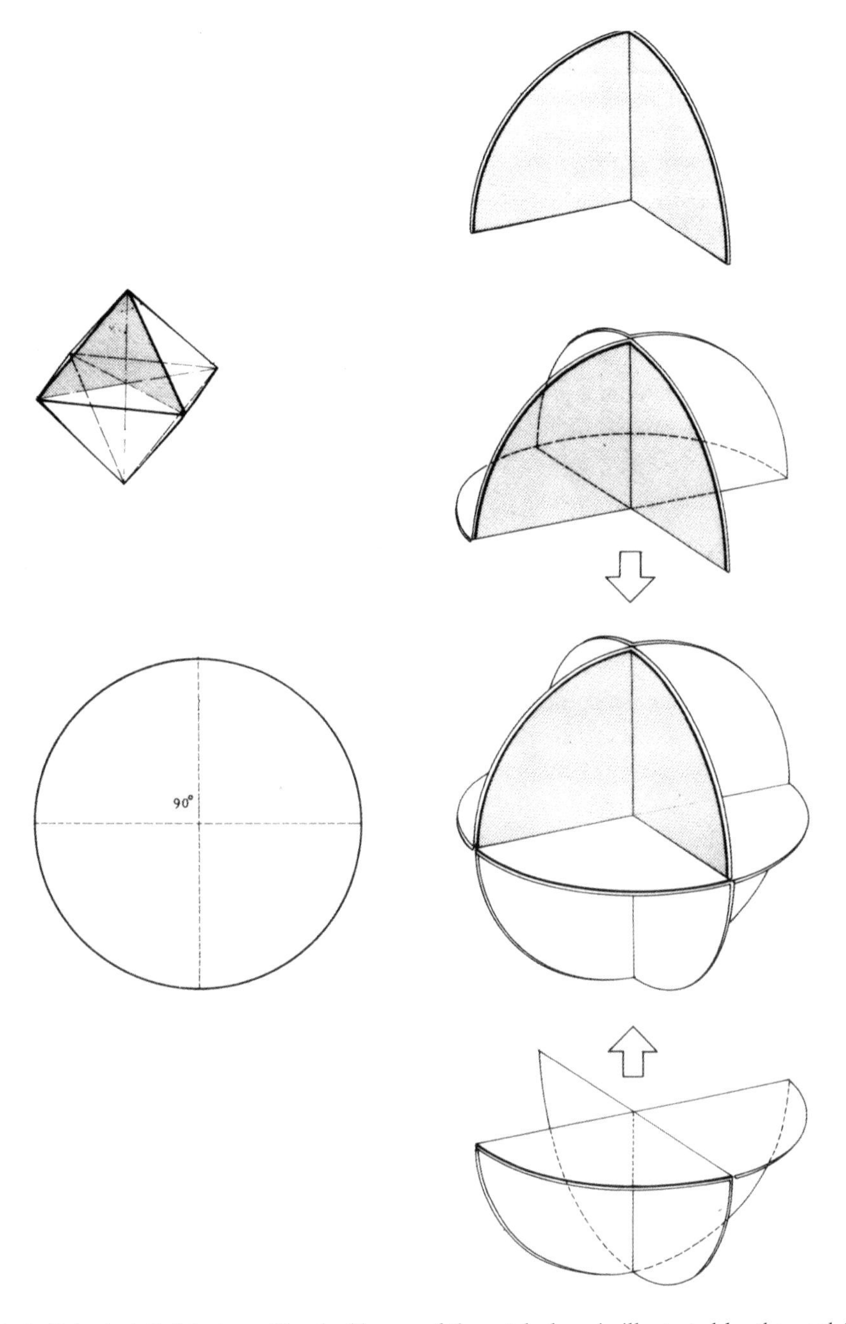

Six-great-circle Spherical Octahedron: The doubleness of the octahedron is illustrated by the need for two sets of three great circles to produce its foldable spherical form.

Four primal yantra shapes, based on mathematical equations from the Sanskrit treatise on mathematics, the Ganita Kaumudi (1356)

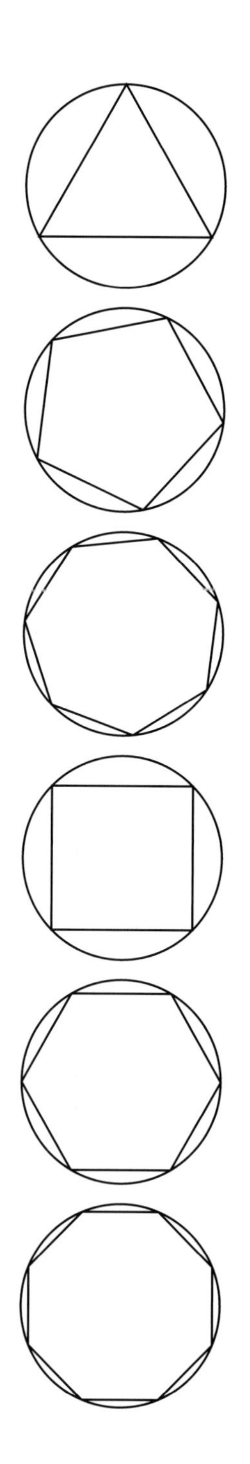

Archetypal shapes based on the division of a circle.

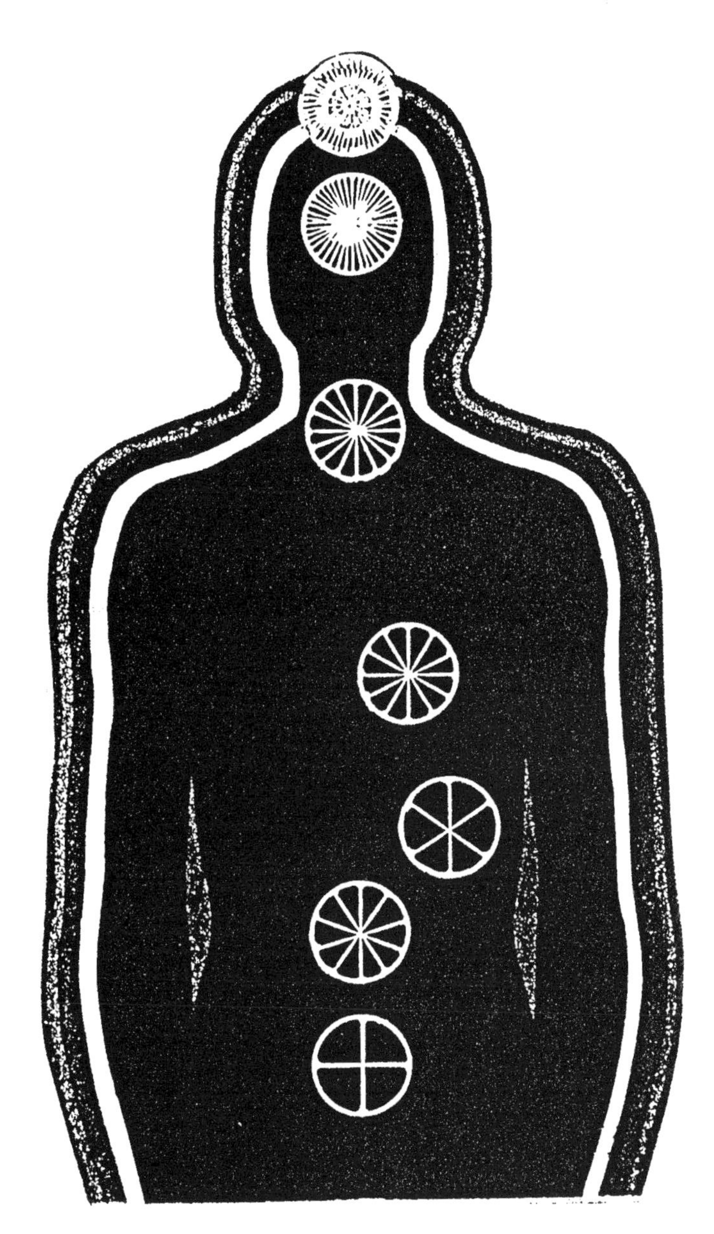

Chakras
(© Janet A. Evans, from C. W. Leadbeater, The Chakras, *Madras/London, 1927.)*

Japanese

Chinese

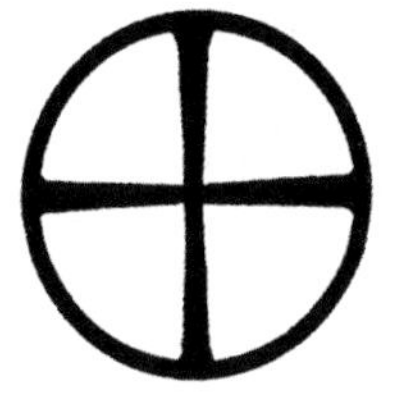

Nordic

7. Frequency – 12-Great Circle Vector Equilibrum

All experiences resolve themselves into discrete angle and frequency patterns. (¶505.03)

Time as Growth

If you have unfolded the WHOLES presented so far, you now have in your hands the basic strategies Nature uses for all her myriad interpatternings. But growing things – be they flowers, crystals, thoughts, or dreams – take *time* to articulate themselves.

In the late 17th Century, Newton still perceived both time and space as absolutes, a hand-me-down assumption from the classical Euclidean worldview. More than two centuries later, Einstein disagreed, postulated four-dimensional space/time relativity, and proposed some of the practical tests which eventually confirmed his intuition. Following on this overthrow of the clockwork universe, the Russian 'truth-seeker' P. D. Ouspensky defined time in his *New Model of the Universe* (1931) as "the dimension of organic growth." His model for the 'shape' of time was a tree.[11]

Fuller's intuition of time is similar, but more precise. *Time* in his synergetic geometry is the dimension of growth out from centers, i.e., from the handful of WHOLE systems we have been toying with in the first four models (spherical vector equilibrium, octahedron, icosahedron, and rhombic dodecahedron). Time is the *frequency* with which such primordial events occur and recur; time begins with this recurrence, *the second.*

To give you some of the flavor of Bucky's own approach to unfolding "nonsimultaneously conceptualizable scenario Universe" from these relatively simple WHOLES, it seems appropriate to cite the Master Shaper himself. He requires of his

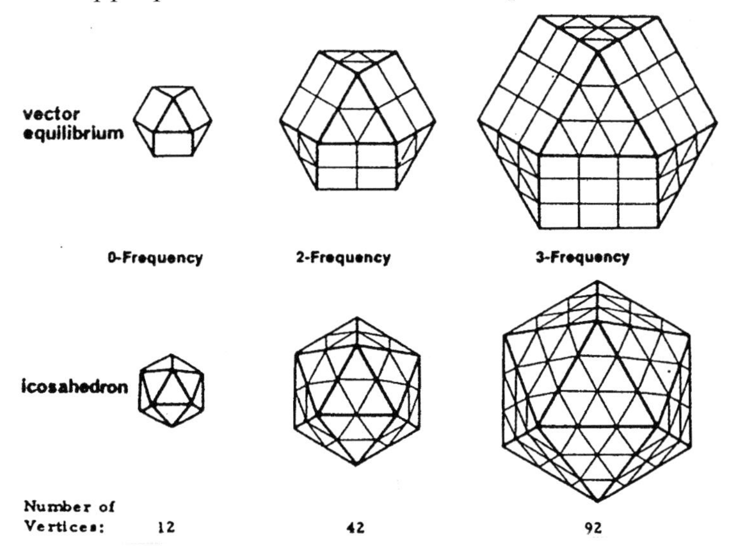

readers nothing less than a radical re-envisioning of everything we take for granted about time and space and all their relations.

Frequency as Time

The Babylonians tried unsuccessfully to reconcile and coordinate time and space with circular-arc degrees, minutes and seconds. The XYZ, c.g.s. metric system accounted time as an exponent. Time was not a unique dimension.

Synergetics is the first [geometry] to introduce the time dimension integrally as the frequency for [WHOLE] systems, which initially are independent of time and size, but when physically realized have both time and size, which are identified in synergetics as the frequency *of the system: the modular subdividing of the primitive, timeless, metaphysical system. (¶1054.70-72)*

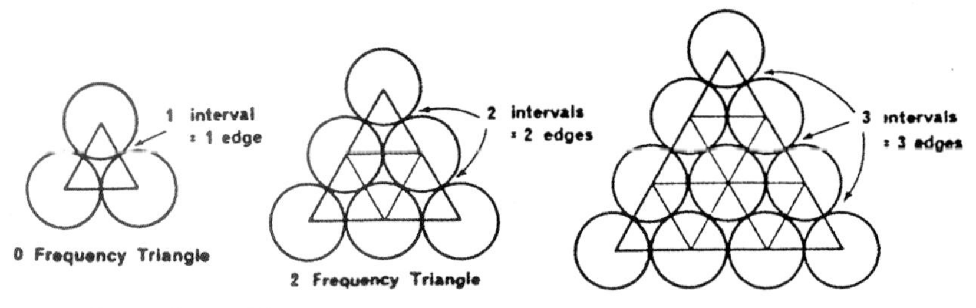

You cannot have time without growthability, which implicitly has a nucleus *from which to grow. We could not have discovered the frequency or time dimensions had we not explored the expansiveness-contractiveness and radiational-gravitational behavior of nuclei [i.e., WHOLES] in pure metaphysical sizeless and timeless principle. (¶1054.72)*

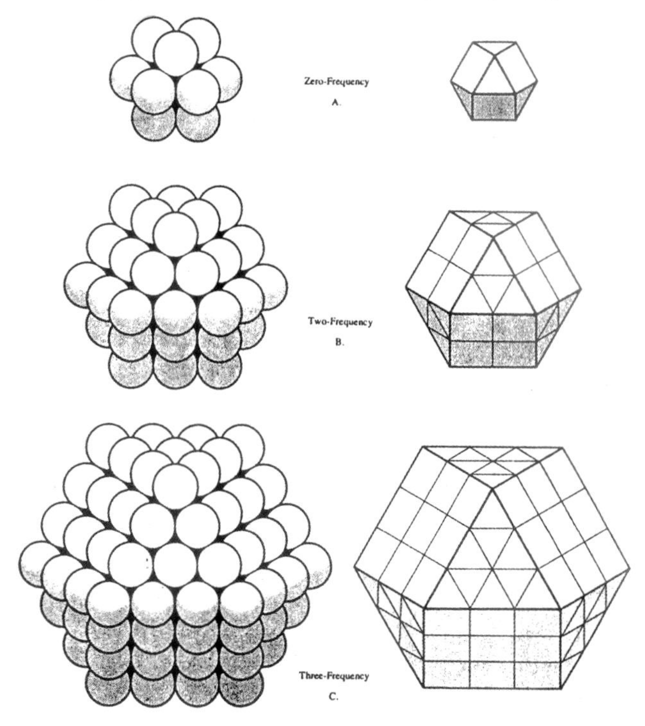

Frequency Shell Growth

Closest packing of spheres around a nuclear sphere always produces vector equilibria of various frequencies; the more spheres, the higher the frequency.

> *As additional layers of equiradius spheres are added it is found that a symmetrical pattern of concentric systems repeats itself. That is, the system of three layers around one sphere, with 92 spheres in the outer layer, begins all over again and repeats itself indefinitely with successively enclosing layers in such a way that the successive layers outside of the 92-sphere layer begin to penetrate the adjacent new nuclear systems. We find then that only the concentric system of spheres within and including the layer of 92 are* unique *and individual systems. (¶413.04)*

This concept of a finite system in universal geometry is directly related, Fuller maintains, to the significance of there being 92 and only 92 regenerative chemical elements. R. Marx, in his *Dymaxion World of R. Buckminster Fuller,* extrapolates:

> *It was no surprise to Fuller when the transuranic elements were developed ... and it was found that these 'elements' disintegrated in split seconds. Fuller describes the transuranics as trans-vector-equilibrium configurations, that is, atomic arrangements in which the radial vectors (the 'explosive' force lines) exceed the circumferential restraints.*[12]

By contrast, platinum, the densest and most durable of metals (it is the international standard for weights and measures) turns out to be a near-perfect high-frequency vector equilibrium. When Fuller was shown the photograph of a single platinum crystal reproduced on p.170, he remarked that the information storage capacity of heavy metals would soon be found to far exceed that of silicon, and, properly utilized, would increase the power and diminish the size of computers dramatically. Twenty-five years later, microchip manufacturing giants Intel and IBM simultaneously announced "one of the biggest advances in transistors in four decades" (*Washington Post*, 27 Jan. 2007), which appears at least partially to confirm Bucky's intuition. Their breakthrough involves using the silvery heavy metal hafnium to regulate the flow of electricity in a new generation of transistors, ensuring that microchips will continue to 'ephemeralize', as Bucky would put it, getting ever smaller and yet more powerful well into the forseeable future. Once you've finished unfolding the Frequency Vector Equilibrium described below, you need only hold the 4/8-way apertures of your completed model up to the photograph to see its striking similarity to the platinum crystal.

Indeed, this is more than similarity; it is identity of structure. Of course there is no platinum in your model, or it would be worth a great deal of money. The atoms of platinum are the white smudges in the photo, which cannot be picked out individually by the field ion microscope. The black spaces in the photo, which delineate the structure you have unfolded, are actually the *empty spaces* where there are no platinum atoms. This is what Fuller means when he speaks of the entirely *meta*-physical pattern integrity of the crystal.

Since this Appendix is just an introduction to Fuller's geometry through his great-circle models, we shall have to refrain from elaborating the very many further correspondences with natural patterns available in Fuller's *Synergetics* volumes. You do know just about all you need to know in order to decide whether you wish to invest the time and effort that will be required to unfold your first frequency WHOLE, the 12-great-circle vector equilibrium outlined at the end of this section. It is perhaps the most elegant, and certainly the most challenging of all the WHOLES. It will take a little time for you to put this WHOLE together, but you don't begrudge it. Time is precisely the point, as Fuller duly reminds us:

> *The only dimension is time, the time dimension being the radial dimension outward from or inward toward any regenerative center, which may always be anywhere, yet characterized by always being the center of system regeneration.*
>
> *The time dimension is frequency.*
>
> *Any point can tune in any other point in Universe. All that is necessary is that they both employ the same frequency ... (¶960.06-08)*

So what you need to unfold a frequency WHOLE is not so much a set of instructions, but a certain attunement to the synergetic principles at work in the WHOLE:

> *In the equanimity model [vector equilibrium], the physical and the metaphysical share the same design. The whole of physical Universe experience is a consequence of our not seeing instantly, which introduces time. As a result of the gamut of relative recall time-lags, the physical is always the imperfect experience, but tantalizingly always ratio-equated with the innate eternal sense of perfection. [¶443.04]*

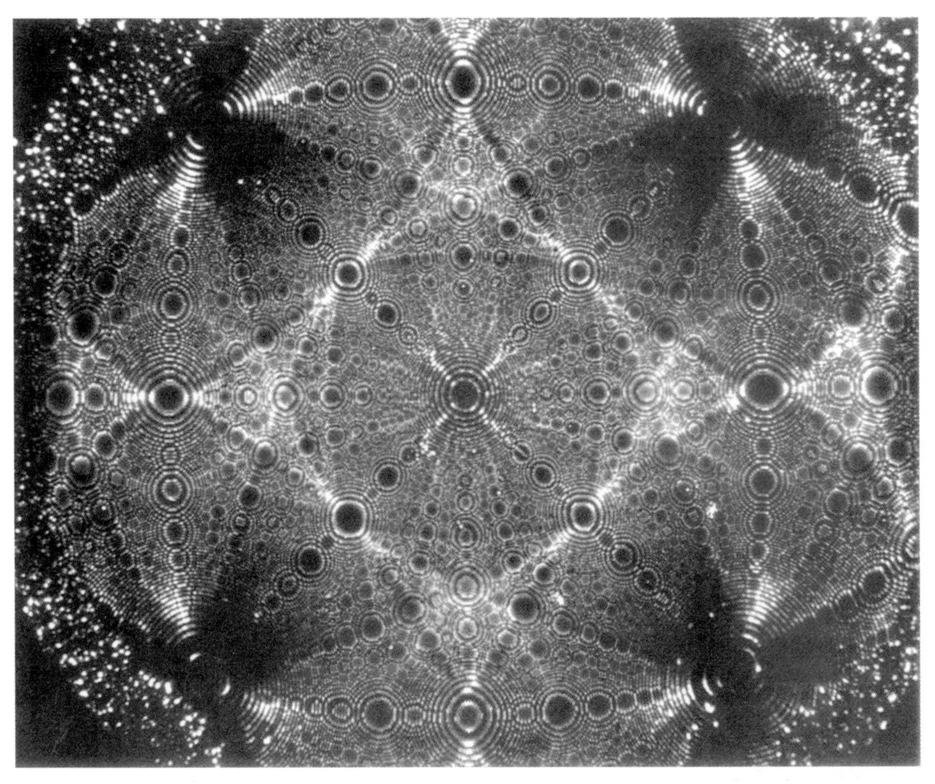

A field ion microscope picture records the actual locations of individual platinum atoms arranged in clusters within the ringlike facets of one crystal of the metal. The atoms – the tiny white dots – were magnified 120,000 times on the negative and here are further enlarged to 500,000 times life size. (Erwin Müller, Platinum crystal, from Photography as a Tool, *New York, 1970.)*

Unfolding the Frequency Vector Equilibrium

12 disks
132 pins
Fold and pin bow-ties.

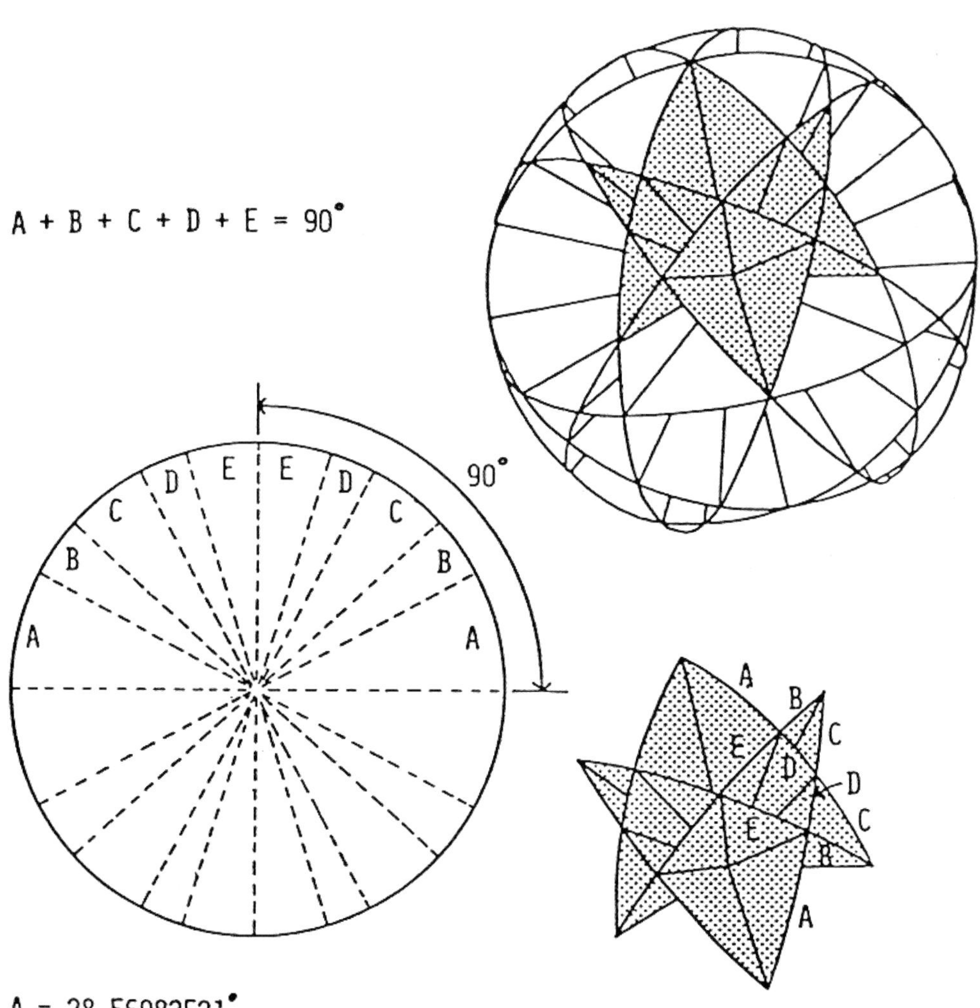

A = 28.56082521°

B = 14.45828792°

C = 19.28632541°

D = 10.67069527°

E = 17.02386618°

8. Dwelling Valve – 10-Great-Circle Icosahedron & Buckyballs

Often circumlocutious, Fuller minces no words on the following point: "The icosahedron is the prime dwelling valve of Universe." Beyond the theoretical underpinnings he gives for the ubiquity of icosahedral shells in Nature, Fuller is best known for developing high frequency icosahedra into a unique form of human dwelling, the geodesic dome.

A 'geodesic' is simply a great-circle arc, but as Fuller spun out many more complex figures from the 'geodesics' in the WHOLES covered so far, the term has come to denote a vast range of domeworks strategies in which the original great circles are not always visible. Geodesic domes can be generated from from spherical tetrahedra, vector equilibria, octahedra, and so forth but Fuller (and Nature too, it seems) has settled on the icosa as the most economical method for enclosing the most volume with the least surface area, and therefore the least expenditure of materials.

American Society for Metals Dome, Cleveland. (© Robert Duchesnay, 1990)

Since physics has found no continuums, we have had to clear up what we mean by a sphere. It is not a surface; it is an aggregate of events in close proximity. [¶1023.11]

The Greeks were comfortable with the idea of perfect and infinite spheres, a notion which allows for many convenient abstractions from 'this' world to the 'ideal' world of Forms, but Fuller insisted that today we can no longer afford to live with the misleading idea of 'solids.'

We find local spherical systems of Universe are definite rather than infinite. ... All spheres consist of a high-frequency constellation of event points, all of which are approximately equidistant from one central point. All the points in the surface of a sphere may be interconnected; they will subdivide the surface of the sphere into an omnitriangulated spherical web matrix. [¶224.07]

Moreover, getting back to our original point about the icosahedron:

All spheres are high-frequency geodesic spheres, i.e., triangular-faceted polyhedra, most frequently icosahedral because the icosasphere is the structurally most economical. [¶985.22]

As a sidelight which reveals the way he never stopped thinking about such elementary things, notice that in *Synergetics* Fuller was already pointing out that 'polyhedra' is in fact a misnomer, because it suggests solid 'faces' (*-hedra*). In *Synergetics 2*, he went so far as to coin a more accurate term:

There are no solids or absolute continuums; ergo, there are no physically demonstrable faces or sides or hedra; ergo, we reidentify the system-conceptioning experiences heretofore spoken of as polyhedra, by the name polyvertexia, the simplest of which is the tetravertex, or 'four-fix' system. (¶986.728)

One way to view frequency modulation is as edge-division of the basic triangles of the system. Classical domeworks exhibit one of two strategies:

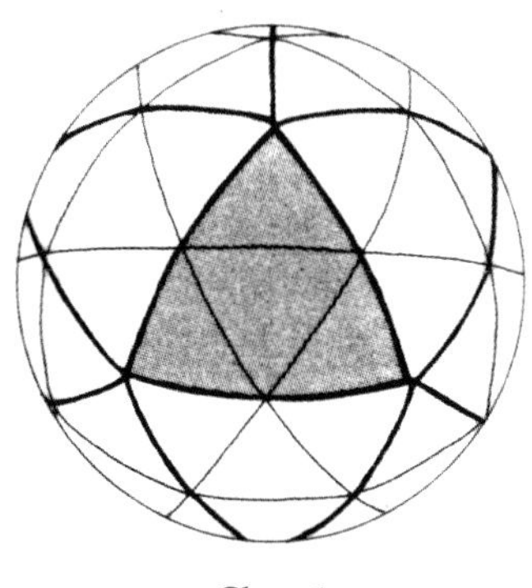

Class 1

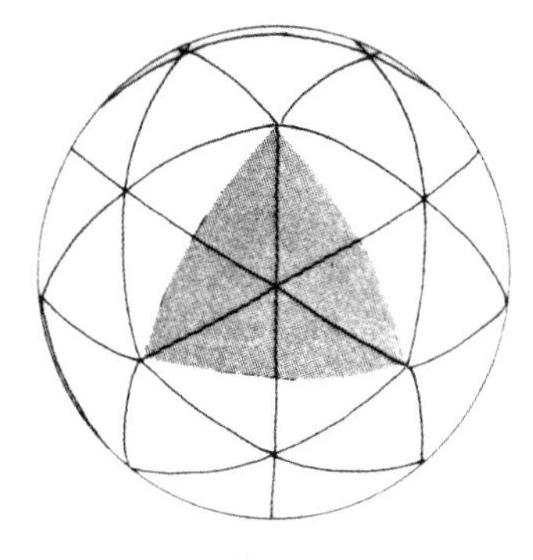

Class 2

These alternatives have differing strengths and weaknesses for various sorts of dwelling functions. *Domebooks I* and *II* are a good review of techniques and complaints, e.g., the importance of making models before building the real thing, the leaky roofs on wooden domes (which expand and contract) unless carefully shingled, etc. Hugh Kenner has also explicated many further dome design options in his *Geodesic Math, and*

How To Use It.[13] In this work, which he says was written for one architect daring enough to try them, Kenner deploys formulæ for designing complex elliptical and egg-shaped domeworks that do not stick to the rigid sphericality Fuller preferred. This should enable the architect or homebuilder to explore uniquely local applications of synergetic architecture ... If you want a window there by the tree, and a door over there where it's sheltered, Kenner gives you the mathematical tools to spin your own sort of geodesic web to suit.

The dome-building industry seems to be undergoing a resurgence in the early 21st Century as aging baby-boomers reclaim a little of the counterculture spirit of the '60s by custom-designing their own dream homes in all sorts of original ways. Double- and even triple-domes are more and more popular today as people look to create unique spaces in iconic places like Berkley or Sedona, or just to reinhabit their suburban neighborhood in Pleasantville.

Nowadays, everybody has heard of Carbon 60, 'Buckminsterfullerene,' so called because it consists of 60 (tetrahedral) carbon atoms ... basically a big soot bubble, a cage-like structure so large it can house other molecules inside it. This surprising third form of carbon, isolated almost simultaneously by Smalley in the US and Kroto in the UK, has fascinated organic chemists from day one, although it has yet to find any practical application. Smalley says ruefully, "Bucky hasn't found a job." He doesn't seem to have asked whether 'Bucky' was looking for a job, or instead disclosing something about the shape(s) the world is in.

Of course it is a great honor for Bucky to have had such a fundamental molecule named after him, even posthumously. It seems to us, however, that the name is just about all the scientists took from Fuller. Any student of Bucky's synergetic geometry will tell you that the soccer-ball diagram conventionally used to depict C60 is not a Fuller figure at all: There are no triangles! 'No triangles, no structure,' Bucky would say. Therefore if you want to see how the 60 tetraheda of 'Buckminsterfullerene' actually fit together, you have to go back and look at Bucky's own geodesic structures.

The WHOLE we present here may be considered the original 'Buckyball,' the ten-great-circle icosa Fuller himself designed long before all the headlines and kudos and conferences about Carbon 60. It is indeed a simpler affair than most of his domes, exhibiting the familiar 'hex/pent' configuration of the most rudimentary Class I geodesics. It is also incredibly tough. Carbon 60, we are told nowadays, may well be the seed-form of carbon chemistry in the Universe at large, and possibly the oldest form of matter altogether. It's your basic stardust, carbon soot cast out of ancient stellar furnaces. As such, it has had to survive all the rigors of billions of years in space before sifting onto the surfaces of planets like our own little Earth, where it might over a few further billions of years evolve into more complex patterns of life – like you and me, for instance.

Now this WHOLE has one feature we've not met with until now: polar asymmetry. It has a positive and a negative pole. Consequently, half the bow-ties must be folded into mirror-images of the other half. This feature may well indicate a difference in electrical potential, a positive and a negative pole – which presumably plays a role in the 'quantum tunneling' exploited by the scientists at Lawrence Livermore Laboratories who recently constructed a functional 'nano-transistor' by sandwiching a single Buckyball between gold electrodes – the first and so far the only nanotechnology component ever successfully tested.[14] Maybe Buckminsterfullerene will one day find 'a job' after all.

Unfolding this WHOLE is guaranteed to make you see stars, 12 of them to be precise, five-pointed beauties dancing all across its surface. If you use duplex cover stock (white on one side, colored on the other) to construct this WHOLE, you are faced with a choice: Do you want to see white stars surrounded by color, or colored stars surrounded by white?

Recall that these figures are the prototypes for Bucky's domes. They raise fundamental questions … Just what sort of ecological niche would you like to construct for yourself?

Here are the specifications Bucky supplied in *Synergetics*. In this sphere, you're on your own.

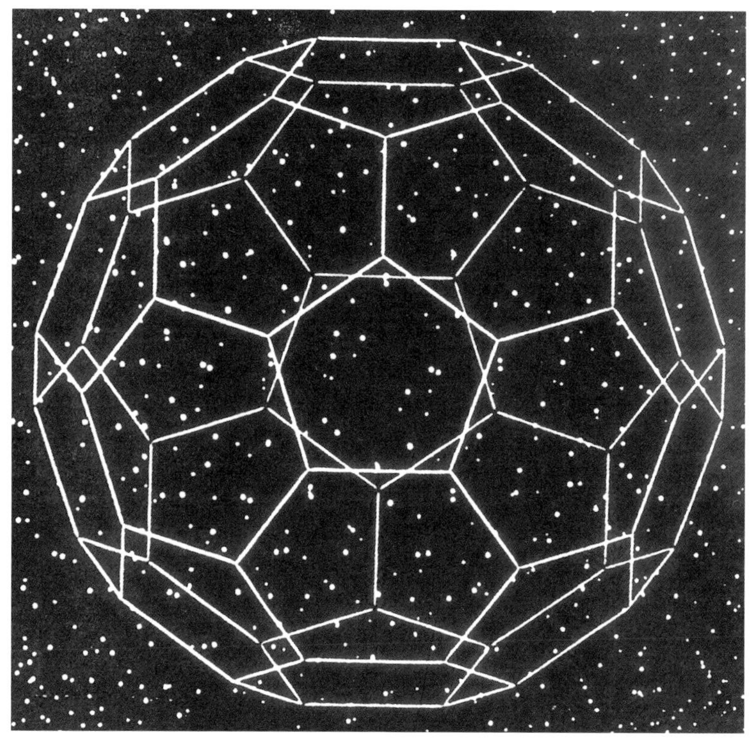

BuckminsterFullerene: C_{60}

A remarkably stable cluster of 60 carbon atoms, formed by vaporising graphite by laser irradiation. Its creators/discoverers suggest this may be the form of carbon carried by diffuse interstellar molecules. (From Nature, *Vol 318 No. 6042 pp14-20 Nov. 1995.)*

Unfolding the Frequency Icosa

10 disks
90 pins
Fold and pin bow ties as shown, half (5) positive and half (5) negative.

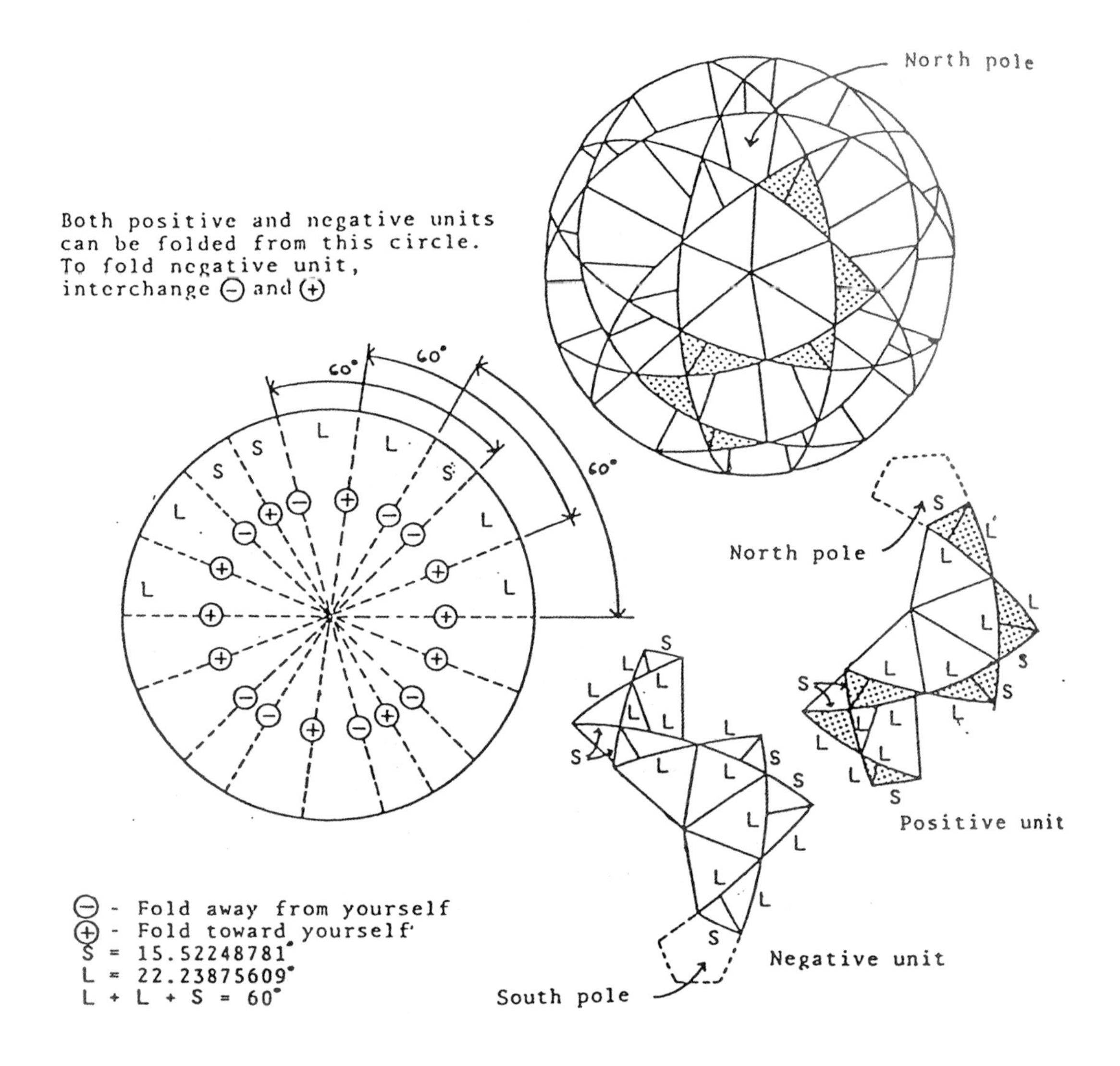

9. Eye of the Beholder – Rhombic Triacontahedron

You see what you look for … This final WHOLE may be described several ways, all of them equally accurate. Depending on your point of view, what we have here can be said to consist entirely of:

- 120 triangles or (volumetric) tetrahedra, the maximum number of like sections into which a sphere can be subdivided; or
- 15 double-edged great circles, each the reassembled version of the 30 disks used to unfold the figure; or
- 30 rhombs, which gives it the name Fuller favors: the rhombic triacontahedron, or
- 20 large triangles or tetrahedra (from the original 0-frequency icosahedron), frequencied by the Class II strategy, making this a 2nd-frequency icosa-dodecahedron, or
- 12 five-pointed star/pentagons, which allows us also to call it a pentagonal dodecahedron.

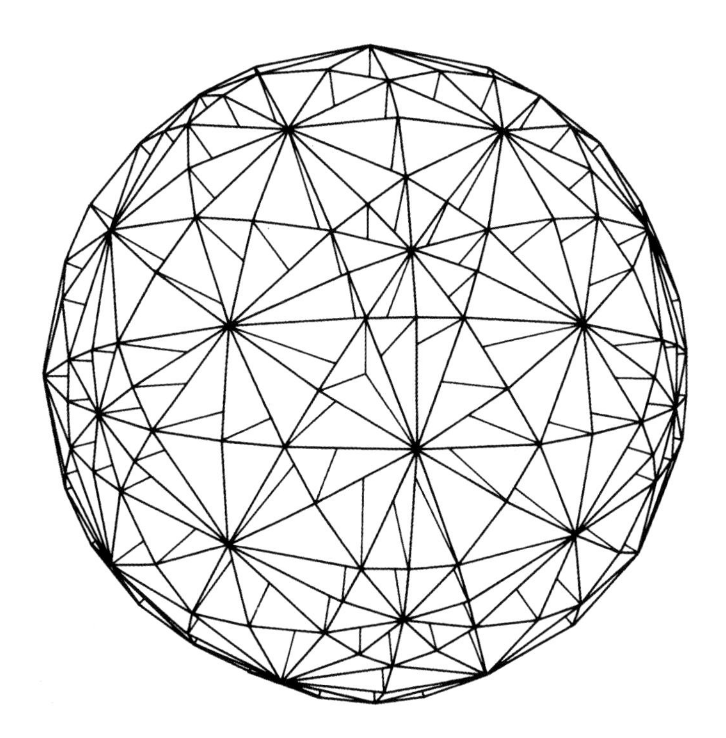

What you 'see' is what you 'get.' Here we cannot help but realize the inadequacy of trying to see the world, at least the energetic-synergetic Universe Fuller lays out for us, from only a single 'perspective.' Genuine pluralism is not a political expedient (until we can convert 'the others' to our point of view), it is the recognition that the whole of reality cannot be reduced to a single 'viewpoint' or common denominator. A pluralistic Universe requires us to see pluralistically, 'in the round,' as it were. As Hugh Kenner mentions, polyhedra are really only conveniences for locating systems of symmetry. All perspectives may be equally valid, but not in isolation from one another. Monofocal vision is (as Fuller spent his entire life demonstrating) mainly a form of culturally conditioned blindness.

Since the 1970s when Fuller articulated his vision in the *Synergetics* volumes, the computerized power to animate and illustrate mathematical theorems has made plausible what seem at first glance to be competing visions of 'the geometry of Nature,' namely fractal geometry and 'chaos' theory. Hugh Kenner has explored the former, Mandelbrot's fractal indices, with reference to Fuller (as well as Ezra Pound), and James Gleick provided a good introduction to the latter two decades ago in his *Chaos*. But the 'self-similarity' of fractal geometry is not synergetic. If the whole is entirely predictable from the part, we are in a different world from Fuller's, perhaps mapping the dissipation and *dis*-integration of systems (the turbulent flow of a river, the erosion of a rocky seashore, the spilling of a can of paint) rather than 'unfolding the wholes' from which flowers, trees, and water droplets take their shape. Chaos theory, in its 'sensitive dependence on initial conditions,' begins with the gritty particularities of experience – the highs and lows of weather patterns, the fluttering of leaves, the dripping of a faucet – and looks behind them for 'strange attractors,' non-linear equations which seem to resemble Fuller's non-linear tensegrities in many ways. Chaos theory focuses on complex, local asymmetries which disclose hidden, global symmetries. Fuller, as we have seen, also sought to bridge the gap between his 'timeless, weightless, metaphysical systems' and the rhythms of order and disorder in everyday life. He devised a series of intermediate geometric steps – mites, sites, couplers, etc. – to bridge the distance between the simple models we have presented here and the complex associations and dissociations of life as we meet it in the ups and downs, ins and outs of the world human beings actually inhabit and experience.

We are neither of us sufficiently sophisticated mathematicians to resolve in some simple formula any disparities between Fuller's version of 'Nature's Coordinate System' and these later computer-enhanced attempts to come to grips with nature's geometry. We may however observe this much: Bucky starts from order and seeks from there to try to account for all the irregularities of the world; the chaos theorists begin with apparent randomness and seek the hidden 'order' behind it. These may turn out to be simply different ways of looking at the oscillations of life, each valid in its own domain. It would be revealing to see Fuller's geometry similarly spun out in computer projections – starting with the triangle unfolding into the tetrahedron, from there into the isotropic matrix, and through that matrix to the full range of WHOLE systems – but this is not our business. Our primary concern here is simply to outline Fuller's synergetic geometry so that you may see for yourself the remarkable correspondences between his synergetic patterns and the sacred geometries of almost every traditional culture. None of those cultures had access to computers, but this did not prevent them from building their temples and cathedrals on these patterns, nor from employing such patterns in symbolic systems intended to bring human culture into meaningful harmony with Nature and, indeed, the Divine.

The final WHOLE we present here is the rhombic triacontahedron, which models the maximum number of like units into which a sphere may be subdivided.[15] Fuller sees in this figure the graphic illustration of Einstein's equation, $E = mc^2$. He called it the "Demass Model," because it seems to present us with the threshold between radiation radiating and matter materializing:

> ... *the difference between* it is matter *and* it is radiation ... *Vastly enlarged, it is the same kind of difference between a soap bubble existing and no longer existing – 'bursting,' we call it – because it reached the critical limits of spontaneously coexistent, cohesive energy as atoms-arrayed-in-liquid molecules and of atoms rearranged in dispersive behavior as gases.*
>
> *This is the generalized critical threshold between* it is *and* it isn't. *(¶986.547)*

Which seems like the proper note on which to conclude this brief synergetics primer.

Unfolding the Rhombic Triacontahedron

30 disks
180 pins

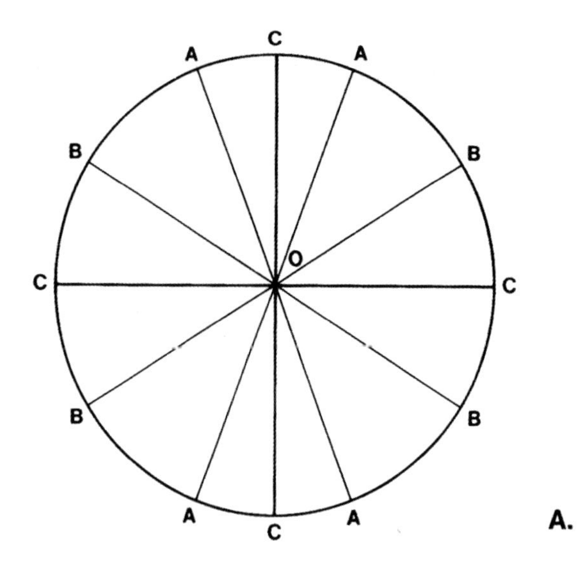

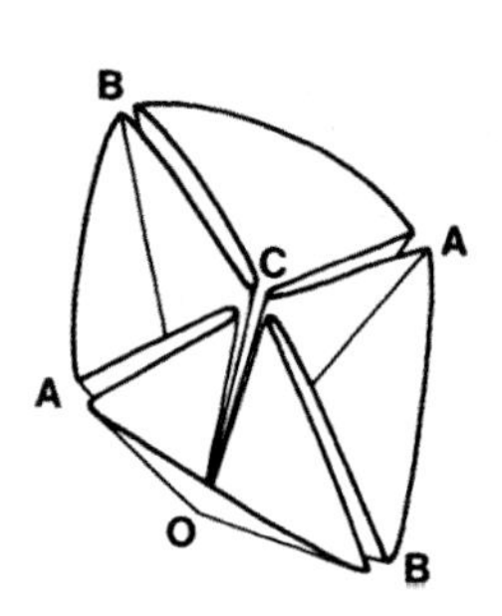

$\angle$ COA = 20.90515745° = arc sin $\left(\sin 18^\circ / \sin 60^\circ\right)$

$\angle$ AOB = 37.37736814° = arc sin $\left(\sin 18^\circ / \sin 60^\circ \sin 36^\circ\right)$

$\angle$ BOC = 31.71747441° = arc sin $\left(\sin 18^\circ / \sin 36^\circ\right)$

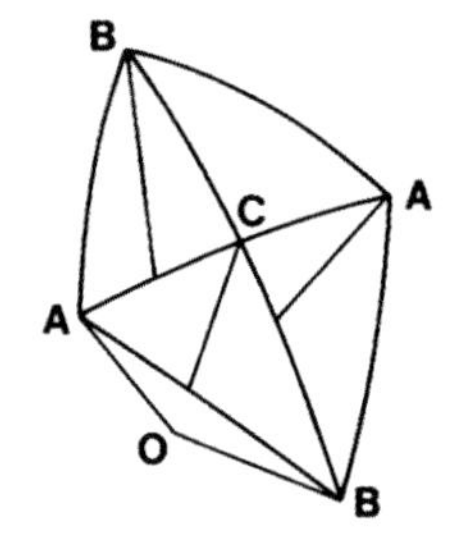

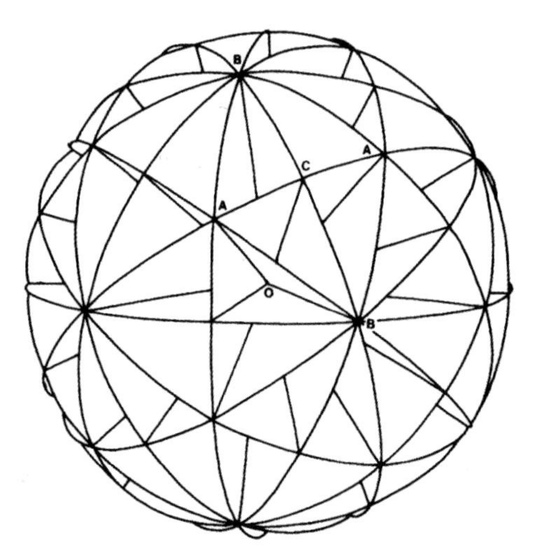

D.

Fuller on Euler: V + F = E + 2

A final resumé may be drawn from Fuller's summary of Euler's topology. In topology, Euler says, in effect, that all visual experiences can be resolved into three unique and irreducible aspects:

- vertices, faces and edges; or
- as unique dimensional abundances:

— points, areas and lines; or

- as structural identifications;

— joints, windows and struts; or

- as we say in synergetics topology:

— crossings, openings and trajectories; or

- the more generalized:

— events, non-events and traceries; or

- more refined as:

— fixes, discontinuities and continuities; or

- in most refined synergetics:

— events, non-events and even inter-relatabilities. (¶1007.22-3, paraphrased)

As an empty set, Euler's equation says that in polyhedra the number of [crossings] plus the number of [openings (holes)] is always equal to the number of [boundaries] plus the number 2 [poles].[16]

Examples:	**V + F = E + 2**
VE	12 + 14 = 24 + 2
ICOSA	30 + 20 = 48 + 2
CUBE	8 + 6 = 12 + 2

Unfolding WHOLES: Practicum
The Medium is not the WHOLE Message

The easiest method of constructing WHOLES is to cut out, score, and fold disks of ordinary construction paper, and then pin them together with bobby pins. Once you've tried a few of these, you will no doubt be impatient to explore other materials. There's more than one way to unfold a WHOLE.

What we have presented here is not so much a series of pretty figures as it is a nascent craft, which could become a genuine vehicle for the creative imagination. Don't be hypnotized by the figures themselves: WHOLES are only a blank multidimensional canvas. Any masterpieces painted on them will be your own. All you need, beyond Fuller's specifications for the figures themselves, is a compass, a straight edge, a little imagination and the right materials. Even scissors are an optional accessory.

A simple way to begin discovering new configurations is by drawing your own mandalic patterns on the disks themselves and watching these re-assemble in four dimensions when you put the model together. Or you could try color-coding the disks to bring out one or another of the patterns latent in the WHOLE.

All the physical 'parts' of the WHOLE are replaceable by other pins, folds or materials. What remains is only the metaphysical integrity of sheer principle. WHOLES are not just material objects; the message is more than the medium in which it is embodied. But just as there is no message without medium, there are no WHOLES *in vacuo*. Here are some materials and strategies which have worked well for us:

Pins

Plain old commercial bobby pins are great for paper WHOLES, and they're flexible enough to tolerate most of the trial-and-error of learning to unfold WHOLES. For a more finished look (tighter vertices, sharper great circle planes), try replacing bobby with cotter pins. We've found that 1" x 1/16" cotter pins work very well for figures from 3" to 12" in diameter. For larger figures, many bigger, fatter and longer cotter pins are available at hardware stores. (Just be sure to avoid the oily ones.)

Materials

WHOLES can be unfolded from almost anything that will take a fold. Each material has advantages and disadvantages. The following are only suggestions, based on experience.

Paper

For paper models, conventional cover stock is probably the sturdiest and most durable, with some of the tougher construction papers running close second. Two-ply papers, with a primary color on one face and white (or a complementary color) on the other, are handy for highlighting various facets of the WHOLE. You can make floppy WHOLES from rice paper, and practically indestructible WHOLES from heavy cardboard, but the optimum material probably lies somewhere between these extremes.

All colored papers tend to fade with time, and very quickly if exposed to direct sunlight. One way to retard this process is to spray finished WHOLES with clear plastic or acrylic or lacquer. (Do this out of doors. Such sprays are always obnoxous, usually toxic, and highly flammable. Their only virtue is that they will probably preserve your handiwork a little longer).

Vinyl

Vinyl, even if quite thin, has many properties which recommend it for the construction of WHOLES. Its springy, rubbery quality makes folding easy, it holds a sharp crease and the colors do not fade. Vinyl WHOLES are semi-translucent; the back side of the figure is visible as a spinning shadow-play. Vinyl is durable, cleans easily, and is readily available. Its main drawback may be its slightly dull and utilitarian appearance; it lacks the bright glossiness of some of the other plastics mentioned below.

Acetate

Almost too brittle a medium to take a fold without cracking, acetate plastics can if carefully handled turn into strikingly beautiful – if rather fragile – WHOLES. Moreover, at art supply stores you can often find acetate sheets of a deep blue or red color bonded to a mirror surface, either gold or silver, from which can be generated WHOLES of astonishing jewel-like elegance.

Polyester

'Mylar' is a trademark for polyester, a species of petroleum-derived plastics from which clear, chrome, colored or matte-finish WHOLES can be unfolded. Mylar is very strong stuff. You cannot tear it, but scored lightly with a compass point it folds easily and sharply. Instead of using scissors, you'll get more perfect disks by putting steel points in both arms of your compass and cutting each disk out as you inscribe it. Mylar picks up fingerprints during construction: use a clean, dry cloth to polish (or wear gloves in the first place!).

We've tried just about every sort of mylar from the 1 and 2 mil. thickness of colored kite material to the 10 and 15 mil. industrial thicknesses. We have no hesitation in recommending 4 mil. polyester as the optimum WHOLES material we've run across so far. The 4 mil. clear has an unearthly transparency about it, like liquid crystal. Assembled as a WHOLE, it also has the remarkable property of reflecting and projecting light simultaneously.

Spun on a thread (loop it under any pin) in strong sunlight, a clear or chrome WHOLE will fill the room with spinning creatures of light – *devas*, they have been called – a purely unpredictable effect of arranging this sort of material in razor-sharp great circles, which act alternately as windows and mirrors. Mylar WHOLES 'pump' light around a room in ways guaranteed to hypnotize cats, dogs, and small children for hours.

You might carry all this one step further than we did and try laser-cut diffraction grating material which, in theory anyway, ought to give you a roomful of rainbow '*devas*', something like the spectral tentacles of Apple's OS X 'Flurry' screensaver writ large. For such experimental materials, only one rule really applies: Try it. Who knows? You might like it.

Advanced Synergetic Origami

You have more freedom than you may suppose in the design of custom-crafted WHOLES. You need not stick exclusively to the ('seven and only seven') intact-great-circle models Buckminster Fuller himself designed. It is possible to unfold quite a few irregular WHOLES as well. Your ingenuity and the limits of your material are about the only constraints.

For example, five disks folded without a continuous diameter into nine equal 40-degree sections will yield a strange and wonderful cross-section of metamorphosis: A 2nd-frequency spherical tetrahedron in the midst of transforming itself into the 3rd-frequency spherical tetrahedron sprouting on the other side of the figure. (see Figure 'SOMEWHOLESOMEWHOLES', p. 137). Or, with six disks scored at equal 45-degree angles, the eight radii provide folds from which one can easily construct a version of what is called 'Kelvin's polyhedron,' with its alternating 4-lobed stars and hexagonal apertures.

Pure symmetry can be tedious. This is why the WHOLES – clear and mirrored, or faced with opposing colors – often turn out to be more interesting as art objects than Bucky's 'static' mass-produced domes. Strung as a mobile on a string, or turned to this peculiar angle or that, the model acquires the added dimension of *change through time*. You may see just a few oblique facets and odd angles, yet you are still able to sense the unseen parts and to intuit the overall pattern of the WHOLE. WHOLES are beautiful for the same reason that a field of early spring daffodils is beautiful: endless intricate variations on a clearly discernable theme.

But if you've read this far, you've got the idea. In model-making as in life, there's no substitute for experience. If the exercises above leave undaunted your ardor for model-making, then 'whole' families of tensegrity spheres (sticks and wire) and geodesics await your attention in Anthony Pugh's *Introduction to Tensegrity* and Hugh Kenner's *Geodesic Math*.[17]

Appendix B

The Root DHṚ and its Branches

Pronounced *dhri*. From the Oxford English Dictionary:

> *Some philologists refer the word to the root of* ferire, *to strike; others compare it with Skt.* dharman *(neut.), holding, position, order, from* dhar, dhṛ, *to hold. The word has been adopted, and is in familiar use in all the Romance and modern Teutonic languages. (p. 1059)*

Quick rundown, from Skeat's *Concise Etymological Dictionary of the English Language:*

(From Latin) DHER, to support – *form-a,* form – conform, deform, form, formula, inform, reform, transform

Also: *firm-us,* firm – affirm, confirm, farm, firm, firmament, infirm

TEN, to stretch –
tend-ere, to extend – attend, contend, distend, extend, intend, intense, intent, ostensible, ostentation, portend, pretend, subtend, superintendent, tend, tender, tendon, tense, tension, tent (1), tenter, toise.

ten-ere, to hold – abstain, appertain, appurtenance, attempt, contain, content, continent, continuous, countenance, countertenor, detain, entertain, impertinent, obtain, pertain, pertinacity, pertinent, purtenance, rein, retain, retinue, sustain, tempt, tenable, tenacious, tenacity, tenant, tenement, tenet, tenon, tenor, tent (2), tentacle, tentative, tenure; cf. tend, lieutenant, maintain.

Also: TRY – to cull, pull out wheat from pulverized chaff; and trotting on track, or just travelling like a train. Perhaps also DRIVE – horse pulling carriage; and DWELL, pulling away, going astray, lingering.

tenu-is, thin – attenuate, extenuate, tenuity; cf. thin.

ATTEND, from French, OF *atendre,* to wait, Latin, *attendere* (pp. *attentus*), to stretch towards, give heed to. Latin *at-* (for *ad*), to; *tendere,* to stretch. From which is derived ATTENTION.

And from Webster –

DRAW, from Skt. *dhrajati,* it moves, sweeps on. That's definitely DHṚ. OS *dragen,* to bear, carry; hence drag, dray. German *tragen,* to pull along. Draw out, draw forth, draw up; in patternmaking, to pull from a mold: draw.

Cf. also DOOR, AS *dor,* Celtic (acc. to Graves) DUIR, the oak, Welsh *derw,* Gaelic *duir,* an oak: not just for strong wood, but the proper wood for midsummer fire-sacrifice of Year-King, Green Man, etc. The *axis mundi* tree, the 'hinge' between seasons, the other face of Janus … (Cf. *Skambha* in the Vedas.) Cf. Trust, truth, troth, tryst, trig: all related via DHṚ and also TREE, a different Sanskrit root, *dru, daru;* but probably, says Eric Partridge in *Origins,* tied via Old English *treowe,* loyal, trusty ("a true friend") and *treow,* loyalty, fidelity, to *treow, treo,* a tree: "as firm and straight as a tree." Trust the truth of the trees and trusses … Note: The Druid, *oak*(tree-)*seer,* bard/shaman, is primordially the seer of DHṚ. (Here also may lie the secret root of all Celtic iconography, based on the vector equilibrium and isotropic vector matrix; cf. Gundesrup Cauldron, Ardagh Chalic, Book of Kells, etc., for examples.)

And is the root STER not derived from DHṚ? Cf. STRUCTURE, *struct-* to build, *-ure;* – once in English, it seems to be "to heap up," i.e., stones for your house. But the root is Skt. *strnoti, strnati,* strews, scatters (cf. strategy, stratum, straw, street, strew, stroma (supporting framework). So it is the notion of supportive structure from strewing, scattering (as seeds are broadcast), a tensional notion of structure – as a net is cast, or a web spread – rather than the compressive (not to add obsessive) piling up of the heaps of stones and wood we tend to live in nowadays.

And thus of course instruct, destroy, construct, etc. Cf. also stretch, strength …

Fuller: Triangle IS structure. TRUSS must be the same root, via Fr. *trousser, tourser (trousseau),* to pack, fasten up. Also trice, trise, to haul up.

TRUE, truth, trow, troth, trust(y) – firm, believable, protective, reliable, also derived from the root (DHṚ). Thus *dharma* carries all its multivocal resonances into the Indo-European linguistic 'family' – still recognizable on farflung branches.

THREAD, THROW, THROUGH derive supposedly from Latin root TER, to bore; yet also TURN, *torn-us,* a lathe; Grk. *tornos;* TORQUE, twist, tour, &c.

NB: May have DHṚ at basis … TRE, TER – tertiary, track, train, trot, drive, drill; 'trick' keeps the pulling, twisting, torquing connection: to bore with a screw ["spiral" = "triangle" – RBF].

And of course: THREE, Sanskrit masc. nom. pl., *trayas,* IDG., *treyas;* thrice: not to say, TRINITY. (Cf. triquetra).

'Suchness,' *Dhamma,* & *Kayanerekowa* (the Iroqouis Great Law): This is 'The Way Things ARE …' etc. One's *dharma,* one's 'duty' (the most common and conventional translation of *dharma*) is thus to uphold, maintain, *the order of things as they are:* the fluxus quo.

Appendix C

Palimpsest

John 1,1-3: A Figurative Reading

Originally ▽ unfolded.

▽ unfolded with ▽,

and ▽ unfolded as ▽.

The selfsame principally unfolded as ▽.

The whole through itself

came to pass,

and outside of this

nothing that has happened

ever happened.

from S. Eastham, *The Radix*, Bern (Lang) 1991, p, 133.

Notes

I. Archaeology of a Vision

1. This famous saying (properly, 'God is an intelligible sphere, whose center is everywhere and whose circumference is nowhere.') actually seems to have originated late in the 12th century with the genial French monk Alanus of Insulis (Alain de Lille), and was preserved in the old German *Book of the Twenty-four Philosophers*. Yet from that time to this it has echoed through so many authors in the West – Giordano Bruno found the idea liberating, while to Pascal it seemed terrifying – that even Emerson wasn't quite sure which 'ghost' was speaking to him.
2. Ralph Waldo Emerson, "Circles," from his first series of *Essays*, opening paragraph; cf. any serviceable edition, e.g., R. W. Emerson, *The Works of Ralph Waldo Emerson*, New York (Tudor) 1930, Vol. 1, p. 193.
3. For the mood lately, cf., e.g., Wendell Berry, "Thoughts in the Presence of Fear," *Orion*, Autumn 2001: "The time will soon come when we will not be able to remember the horrors of September 11 without remembering also the unquestioning technological and economic optimism that ended on that day."
4. See R. Buckminster Fuller, "A Comprehensive Anticipatory Design Science," in his *No More Secondhand God and Other Writings*, New York (Doubleday/Anchor) 1963/1971, pp. 75-104.
5. See R. Buckminster Fuller, *Synergetics – Explorations in the Geometry of Thinking*, E. J. Applewhite, Ed., New York (Macmillan) and London (Collier) 1975, pp. 13-17; and R.B. Fuller, with Kiyoshi Kuromiya, *Critical Path*, New York (St. Martin's Press) 1981, pp. 123-60. For a concise summary of Fuller's 'comprehensive anticipatory design science,' see Amy C. Edmonson, *A Fuller Explanation: The Synergetic Geometry of R. Buckminster Fuller*, Boston/Basel/Stuttgart (Birkhäuser) 1987, pp. 258-69.
6. Cf., e.g., E. J. Applewhite, *Cosmic Fishing – An account of writing* Synergetics *with Buckminster Fuller*, New York (Macmillan) 1977, p. 143: "I know that the whole structure of Fuller's cosmos is a poetic one of vast harmony and subtlety."
7. *Science*, Vol. 254, 20 December 1991, pp. 1705-07. Cf. also R. F. Curl & R. E. Smalley, "Fullerenes," in *Scientific American*, Vol. 265, No. 4, October 1991, pp. 54-63, for the discovery of this unexpected third form of carbon.
8. Marshall McLuhan, cover note to Robert Snyder's video, *The World of Buckminster Fuller*, Masters & Masterwords, Pacific Palisades, CA, 1971.
9. Joachim Krause & Claude Lichtenstein, Eds., *Your Private Sky – R. Buckminster Fuller: The Art of Design Science*, Baden, Switzerland (Lars Müller) 1999.
10. "Extraterrestrial Helium Trapped in Fullerenes in the Sudbury Impact Structure," J. Bada, L. Becker, R. Poreda, *Science*, Vol. 272, 12 April 1996, pp. 249-52. A similar case of ancient gases found in Buckyballs may reveal a decidedly less friendly aspect of the Universe; as reported in yet another *Science* magazine (Vol. 302, 21 November 2003),

material preserved in Buckyballs seems to confirm an asteroid collision marking the drastic Permian-Triassic extinction of 251 million years ago, when 90% of the Earth's sea life, and 70% of land-based life, abruptly disappeared.

11. Epigraph, R. B. Fuller, *Intuition: A Metaphysical Mosaic,* New York (Doubleday) 1971; reissued: San Luis Obispo, CA (Impact) 1983. The tale of how Pound came to write this epigraph is told in Buckminster Fuller & Anwar Dil, *Humans in Universe,* New York (Mouton) 1983, pp. 20-30.
12. R. B. Fuller, *Intuition: A Metaphysical Mosaic, op. cit.,* pp. 101-211.
13. From Ralph Waldo Emerson, "The American Scholar," in Stephen E. Whicher, Ed., *Selections from Ralph Waldo Emerson,* Boston (Houghton Mifflin/Riverside) 1957, pp. 79-80.
14. R. B. Fuller, Letter to his sister Rosamund Fuller, 13 June 1928, *4D Time Lock,* Privately Printed 1929; Carbondale, IL (Biotechnic Press) 1970, 1972, p. 79.
15. R. W. Emerson, "Beauty," in *The Works of Ralph Waldo Emerson, op. cit.,* Vol. 3, *Conduct of Life,* p. 193.
16. Cf. R. Buckminster Fuller, *Earth, Inc.,* New York (Anchor/Doubleday) 1973, "The Leonardo Type," pp. 27-72, where Bucky himself expands on the analogy with Leonardo; as well as, e.g., the massive collection of Leonardo material for the 1938 Milano Exhibition in *Leonardo Da Vinci,* Novarro, Italia (Istituto Geographica De Agostini) 1938, 1967, English edition, New York (Reynal & Co.) 1967, which includes pertinent articles by G. Gentile, "Leonardo's Thought," pp. 163-74; G. Chierici, "Dome Architecture," pp. 233-38; C. Baroni, "Leonardo As Architect," pp. 23-60; and A. Uccelli, "The Science of Structures," pp. 261-74.
17. Martin Kemp, *Leonardo da Vinci,* London (Orion) 1988, p. 85.
18. Perhaps Buckminster Fuller's best known book is still his *Operating Manual for Spaceship Earth,* New York (Dutton) 1963, 1971, 1978.
19. Cf. Alexis de Tocqueville, *Democracy in America,* Henry Reeve, trans. (1839), New York (Bantam Books) 2000, Vol. II, Section 2, Chapter VIII, "How the Americans Combat Individualism by the Principle of Self-Interest Rightly Understood."
20. ibid, II, 2, IV, "That the Americans Combat the Effects of Individualism by Free Institutions."
21. Cf. Hector Garcia's current initiatives in the area of 'cultural complementarity,' available online at http://www.us.initiativesofchange.org/stories.html.
22. Eugen Rosenstock-Huessy, *Out of Revolution – Autobiography of Western Man,* Norwich, VT (Argo Books) 1969, p. 683.
23. Andrew Delbanco, *The Real American Dream,* Cambridge, MA (Harvard UP) 1999.
24. Emerson, *Letter* of September 30 (?), 1842, in *Selections from Ralph Waldo Emerson,* S. E. Whicher, Ed., *op.cit.,* p. 215.
25. Cf. Ken Burns' video, *The Shakers* (1985), broadcast on PBS as *Hands to Work, Hearts to God,* that famous slogan originating with Ann Lee herself.
26. Leonard Mendelssohn, public lecture, "Freaks," Shaker Village, Canterbury, NH, 14 April 1988. For historical overview, cf., e.g., E. D. Andrews, *The People Called Shakers,* New York (Dover) 1953, 1963; for the Shakers' remarkable spirituality, R.E. Whitson, Ed., *The Shakers – Two Centuries of Spiritual Reflection,* New York (Paulist) 1983; and for architectural sidelights, C. E. Robinson, *The Shakers & Their Homes,* Canterbury, NH (Shaker Village, Inc.) 1976.
27. Smith, *A New Age Now Begins,* New York (McGraw-Hill) 1976, p. 1172.
28. Cf. the profound and evocative contemporary "Mohawk Welcoming Ceremony" led by Sakokwenionkwas (Tom Porter), in the Conference Proceedings volume, *Living With the Earth,* Montréal (Intercultural Institute of Montréal) 1992, pp. 17-26. Cf. also E. Tooker, Ed., *Native North American Spirituality of the Eastern Woodlands,* New York (Paulist) 1979,

"Iroquois Ceremonials," pp. 268-81.

29. Cf. Robert Vachon's groundbreaking series on "The Mohawk Nation" in *INTERculture,* bilingual Journal of the Intercultural Institute of Montréal, issues #113, "The Mohawk Nation and its Communities: Some Basic Sociological Facts" (Fall, 1991), #114, "Western and Mohawk Political Cultures: A Study in Contrasts" (Winter, 1992), #118, "The Mohawk Dynamics of Peace: The People of the Great Peace" (Winter 1993), and #121, "The Mohawk Nation: Its Seven Communities. A Brief History" (Fall, 1993). Undertaken in collaboration with Mohawk elders, this series of articles amounts to an alternative history of North America from the Iroquois' perspective.
30. Says Robert Vachon: "It is very difficult to synthesize in a paragraph the complex events of the final decades of the twentieth century. The revival of the Traditional Longhouse Mohawk communities started in the 1960s with the establishment of the North American Indian Travelling College, free border crossing, and *Akwesasne Notes*. Then, in the 1980s, there were the very dangerous Racquette Point events, followed in the 1990s by the Oka crisis and the Mercier Bridge events in Montréal. The revival had problems not only with Western White government (U.S. and Canada) but with some of the band councils and their Indian police, and finally between the Traditionalists themselves: those faithful to the Tradition and Condoled Chiefs, and those who went capitalistic (casinos and cigarettes) or who formed the Warrior Society (without the backing of the condoled chiefs, *rotiyaner*), both of which went against the sacred traditions and tried to take over and present themselves falsely as the Longhouse and Confederacy. Elder Tom Porter, already in the 1960s, dreamed of returning to the Mohawk homeland, but only got permission from the Confederacy in the '90s to do so. The 15 Mohawks who moved there with him did not stay. He did and is still there, going strong. He has the encouragement of the Confederacy and traditional Mohawks, but the latter are not ready as yet to leave their village homes and join the Clean Pot permanently. The Clean Pot is doing very well but only a few Mohawks are there full time. What about the future? Tom Porter doesn't give up his dream. He is preparing the way for the seven generations to come." *(Letter to the Author; 17 Dec. 2003)*
31. R. W. Emerson, *The Works of Ralph Waldo Emerson, op.cit,* Vol. 3, p.141.
32. *synergoi tou theou,* God's co-workers, in I Cor. 3:9. Cf. also Rom. 8:28, *panta synergei eis agathon,* all things work together for good (to those that love God, etc.).
33. R. Buckminster Fuller & E. J. Applewhite, *Synergetics, op. cit.,* ¶101.01, p. 3.
34. Cf. Antonio Favaro, *Galileo Galilei: Pensieri,* Florence, 1949.
35. R. Buckminster Fuller, *Utopia or Oblivion,* New York (Overlook Press) 1969, p. 76.
36. Cf. E. J. Applewhite, *Cosmic Fishing, op. cit.*
37. Cf. Hugh Kenner, *Bucky, A Guided Tour of Buckminster Fuller,* New York (Morrow) 1973, out of print but still the most readable and comprehensive introduction to Fuller and his work.
38. R. Buckminster Fuller, *Critical Path, op. cit.* pp. 169-70; 172.
39. Cf. Raymond E. Fitch, *The Poison Sky – Myth and Apocalypse in Ruskin,* Athens, OH (Ohio University Press) 1986.
40. Cf. Jon Jeavons, *How to Grow More Vegetables,* Berkeley (Ten Speed Press) Fifth Edition, 1995; and also the video *Circle of Plenty,* Bullfrog Films, 1987, on Jeavons' Willits farm utilizing the methods pioneered in those early years by Chadwick at UCSC's Garden Project.
41. Mike Davis, *City of Quartz – Excavating the Future in Los Angeles,* New York (Vintage) 1992.
42. Cf. R. Buckminster Fuller, *Critical Path, op. cit.,* Introduction, "Twilight of the World's Power Structures," pp. xvii-xxxviii.
43. Slightly abridged from R. Buckminster Fuller, *Synergetics – Explorations in the Geometry of*

Thinking, E. J. Applewhite, Ed., *op. cit.,* ¶301.10, "Definition: Universe," p. 81.

44. R. Buckminster Fuller, *Intuition,* San Luis Obispo, CA (Impact; Second Edition) 1983, "Brain & Mind," p. 173.

II Refractions

1. R. Buckminster Fuller & Robert Marx, *The Dymaxion World of R. Buckminster Fuller,* New York (Anchor/Doubleday) 1973, p. 83.
2. ibid.
3. Hugh Kenner, *Bucky – A Guided Tour, op. cit.,* p. 206.
4. R. Buckminster Fuller, *Ideas & Integrities,* New York (Macmillan/Collier) 1963, final page (308).
5. Cf. Krausse & Lichtenstein, *Your Private Sky, op. cit.,* "How to Make the World Work," pp. 11-19.
6. R. W. Emerson, "The Adirondacks," in *The Works of Ralph Waldo Emerson, op. cit.,* Vol. 4, pp. 240-50.
7. Cf. Lewis Mumford, *The Golden Day – A Study in American Literature and Culture,* New York (Dover) 1926, reissued 1953.
8. Lewis Mumford, *The Story of Utopias,* New York, 1922.
9. Lewis Mumford, *Sticks and Stones,* New York, 1924; reissued: New York (Dover) 1956.
10. Lewis Mumford, *The City in History: Its Origins, Its Transformations, and its Prospects,* New York (Harcourt, Brace & World) 1961.
11. Lewis Mumford, *Sticks and Stones, op.cit,* p. 111.
12. Lewis Mumford, *Sticks and Stones, op.cit.,* p. 93.
13. Lewis Mumford, *Findings and Keepings,* London (Secker & Warburg) 1975, pp. 29-36.
14. Lewis Mumford, *Technics and Civilization,* New York (Harcourt, Brace & World) 1934.
15. Lewis Mumford, *The Myth of the Machine, Volume 1: Technics and Human Development, New York & London* (Harcourt Brace Jovanovich) 1967 and *Volume 2: The Pentagon of Power* New York & London (Harcourt Brace Jovanovich) 1970.
16. ibid., Vol. 2, p. 56.
17. Of course the most influential US architect in the mid-twentieth century was Frank Lloyd Wright, who not only used a capital 'B' to invoke a transcendental Beauty in his writiings, but explicitly made the quest for such Beauty the aim of his work – indeed, according to Wright, of all 'Art.'
18. Lewis Mumford, *Findings and Keepings, op.cit.,* p. 373.
19. Lewis Mumford, *Myth of the Machine, op.cit.,* Vol. 2, Illustrations 14-15, pp. 148-49.
20. ibid.
21. Lewis Mumford, *The Conduct of Life,* New York (Harcourt Brace Jovanovich) 1951, 1970, pp. 216-43.
22. Lewis Mumford, *Interpretations and Forecasts, 1922-1972,* New York (Harvest/Harcourt Brace Jovanovich) 1973, pp. 487-96.
23. Fuller, *Critical Path, op.cit.,* pp. 225-26.
24. A good start to such a 'dialogue' was made by William Kuhns, *The Post-Industrial Prophets – Interpretations of Technology,* New York (Harper/Colophon) 1973.
25. Lewis Mumford, *The Conduct of Life, op. cit.,* p. 57.
26. ibid., p. 90.
27. R. Buckminster Fuller, *Intuition – A Metaphysical Mosaic, op. cit.,* pp. 23-4.
28. Lewis Mumford, *My Works & Days: A Personal Chronicle,* New York & London (Harcourt, Brace, Jovanovich) 1979, p. 79.

III Reflections

1. Cf. Scott Eastham, *The Media Matrix – Deepening the Context of Communication Studies,* New York/Oxford (University Press of America) 1990.
2. J. Krause & C. Lichtenstein, Eds., *Your Private Sky, op. cit.* p. 16. Cf. also R. B. Fuller, *Synergetics, op. cit.,* ¶460.00 Jitterbug: "Symmetrical Contraction of Vector Equilibrium," pp. 190-96.
3. R. Buckminster Fuller & E.J. Applewhite, *Synergetics, op.cit.,* ¶440.00, "Zero Model," p. 156.
4. Cf. W. K. C. Guthrie, *A History of Greek Philosophy,* Vol. I: *The Earlier Presocratics and the Pythagoreans,* London/New York (Cambridge) 1962, IV, "Pythagoras and the Pythagoreans," pp. 146-336.
5. Cf. Hamilton, E., & Cairns, H., Eds., *Plato: The Collected Dialogues,* Princeton (Bollingen/Pantheon) 1961, pp. 1162-3.
6. On the astonishing range of figurative possibilities evoked by the archetype of the feminine – from cave and vessel to home and dome –, cf. Erich Neumann, *The Great Mother – An Analysis of the Archetype,* Princeton (Bollingen) 1955, as well as his insightful studies of art and artists, e.g., *The Archetypal World of Henry Moore,* published for the Bollingen Foundation, New York (Pantheon) 1959; *Art and the Creative Unconscious,* Princeton (Bollingen) 1959, etc.
7. Hamilton, E., & Cairns, H., Eds., *Plato: The Collected Dialogues,* op. cit., p. 1166. Cf. also Boethius, *The Consolation of Philosophy,* Richard Green, trans., New York (Bobbs-Merrill) 1962, Book III, Poem 9, for an extremely influential early Christian version of this vision from the *Timaeus*: "Oh God, Maker of heaven and earth, Who govern the world with eternal reason, and at your command time passes from the beginning. You place all things in motion, though you yourself are without change. No external cause impels You to make this work from chaotic matter. Rather it was the form of the highest good, existing within You without envy, which caused You to fashion all things according to the eternal exemplar. You who are most beautiful produce the beautiful world from your divine mind and, forming it in your image, You order the perfect parts in the perfect whole. ... You release the world-soul throughout the harmonious parts of the universe as your surrogate, threefold in its operations, to give motion to all things. [*Tu triplicis medium naturae cuncta moventem / connectens animam per sonsona membra resolvis.*] That soul, thus divided, pursues its revolving course in two circles, and, returning to itself, embraces the profound mind and transforms heaven in its own image." (p. 60.]
8. Cf. *Pseudo-Dionysius – The Complete Works,* Colm Luibheid & Paul Rorem, trans., New York (Paulist) 1987, which at long last presents a convincing English translation of "The Celestial Hierarchy," pp. 143-191. Cf. also Paul Rorem's comprehensive attempt to trace the vast and pervasive influence of these seminal works in P. Rorem, *Pseudo-Dionysius – A Commentary on the Texts and an Introduction to Their Influence,* Oxford (Oxford University Press) 1993.
9. Cf. Martin Buber, *Good and Evil,* New York (Scribners) 1952, 1953, "Imagination and Impulse," pp. 90-97.
10. ibid., "The Primal Principles," pp. 99-106.
11. Ezra Pound, *Gaudier-Brzeska: A Memoir,* New York (New Directions) 1961, p. 92. But the intuition here is age-old; cf., e.g., Nicolas Cusanus, *Of Learned Ignorance,* London (Routledge & Kegan Paul) 1954, the first book of which is an attempt by this formidable Renaissance Neoplatonist to offer his own geometric proof of the Trinity.
12. Cf. Matthew Fox & Rupert Sheldrake, *The Physics of Angels – Exploring the Realm Where Science and Spirit Meet*, San Francisco (Harper & Row) 1996. But see also major works in the Neo-Platonic tradition derived directly from the trinitarian intuition of the Pseudo-Dionysius, like Scotus Eriugena's *De Divisione Naturæ (The Structure of Nature),* or St.

Bonaventure's *Hexæmeron* (see Etienne Gilson's careful redaction in *The Philosophy of Saint Bonaventure,* I. Trethowan & F.J. Sheed, trans., London (Sheed & Ward) 1938, especially pp. 267-70), etc.

13. Martin Kemp et al., *Leonardo de Vinci, op.cit.*, pp. 184, 233.
14. Cf. H. M. S. Coxeter, "Mathematical Models," in *Encyclopædia Brittanica,* 1958 Edition, Vol. 15, pp. 73-5, including excellent illustrations: "A knowledge of plane geometry acquired without any reference to models may be said to flatten out the mind and to engender habits of thought which make it difficult at a later stage of mathematical education to explore space of three dimensions. Some early works of Euclid had diagrams intended to be cut and folded, and a work by Cowley of 1752, *New and Methodical Explanations of the Elements of Geometry*, included pieces of carboard for the building up of various models."
15. Indeed, as William Anderson's *Green Man – The Archetype of our Oneness With the Earth,* London/San Francisco (HarperCollins) 1990 underscores, the tradition of the dying/reborn Green Man lately undergoing resurgence – along with all the attendant 'pagan' calendar festivals – not only harks back to Goddess religion (the Green Man is her son/lover) and women's spirituality, but marks as well a re-attunement to this sacred marriage of Sky and Earth.
16. Cf. F. C. Conybeare, trans., *Philostratus' Life of Apollonius of Tyana,* New York (Macmillan) 1912, Book III, 34.
17. John Gribbin, *Stardust,* London (Penguin) 2000.
18. Cf., e.g., Fred Hoyle, Geoffrey Burbidge & Jayant Narlikar, *A Different Approach to Cosmology,* London (Cambridge UP) 2000; Chandra Wickramasinghe, *Cosmic Dragons – Life and Death on our Planet,* London (Souvenir) 2001; etc.
19. Cf. Lynn Margulis, *Symbiotic Planet (A New Look at Evolution),* New York (Perseus/Basic Books) 1998.
20. Cf. R. Buckminster Fuller & E.J. Applewhite, *Synergetics, op. cit.*, ¶250.30, "Remoteness of Synergetics Vocabulary," pp. 70-1.
21. Jan Christiaan Smuts, *Holism and Evolution,* New York (Macmillan) 1926, pp. 85-86; 116-117. Note especially Ch. V, "General Concept of Holism," e.g., "Let me conclude with a word on nomenclature ... According to the view expressed in this [book] the whole in each individual case is the centre and creative source of reality. It is the real factor from which the rest in each case follows. But there is an infinity of such wholes comprising all the grades of existence in the universe; and it becomes necessary to have a general term which will include and cover all wholes as such under one concept. For this the term Holism has been coined; Holism thus comprises all wholes in the universe. We speak of matter as including all particles of matter in the universe: in the same way we shall speak of Holism as including all wholes which are the ultimate creative centres of reality in the world." (p. 116)
22. Cf. e.g., Arthur Koestler, *Janus – A Summing Up*, New York (Vintage) 1978, "The Holarchy," pp. 23-56.
23. José A. Argüelles, *The Transformative Vision – Reflections on the Nature and History of Human Expression*, Boulder & London (Shambhala) 1975, p. 25.
24. "There is a Geometry of Art as there is a Geometry of Life, and ... they happen to be the same," says Matila Ghyka in his classic *The Geometry of Art and Life,* New York (Dover) 1946, 1977; cf. also Darcy W. Thompson, *On Growth & Form*, New York (Cambridge) 1961 (reprint); H. E. Huntley, *The Divine Proportion,* New York (Dover) 1968; Mario Livio, *The Golden Ratio,* London (Hodder/Review) 2002; Peter Pearce, *Pattern is a Design Strategy in Nature*, Boston (MIT Press) 1979; and similar works.
25. Figure from the Karachi Museum (SCALA from Art Resource).
26. An extensive (if sometimes needlessly obscure, even occult) literature suggests itself

here. Among some of the more worthwhile efforts along these lines: Robert Lawlor, *Sacred Geometry – Philosophy and Practice,* New York (Crossroad) and London (Thames & Hudson) 1982; Keith Critchlow, "The Soul as Sphere and Androgyne," in *Parabola,* Vol. III, No. 4, November 1978, pp. 34-43; Dane Rudhyar, *An Astrological Mandala: The Cycle of Transformations & Its 360 Phases,* New York (Vintage/Random House) 1973; Alice Bailey, *Esoteric Astrology,* New York (Lucis) 1951; Lawrence Blair, *Rhythms of Vision, The Changing Patterns of Belief,* New York (Schoken Books) 1975; Agehananda Bharati, *The Tantric Tradition* (New York (Doubleday/ Anchor) 1970; C. W. Leadbeater, *The Chakras,* London (Theosophical Publishing House) 1927; Nathaniel Harris, *Rugs & Carpets of the Orient,* New York (Hamlyn) 1977; Gareth Knight, *A Practical Guide to Qabalistic Symbolism,* London (Helios) 1965 and New York (Samuel Weiser) 1978; Fritz Meier, "The Mystery of the Ka'ba: Symbol and Reality in Islamic Mysticism" (1944), in Joseph Campbell, Ed., *The Mysteries, Papers from the Eranos Yearbooks,* Princeton (Bollingen) 1955, pp. 149-68; Thomas Fawcett, *The Symbolic Language of Religion,* Minneapolis (Augsburg) 1971; and, solely for its 151 illustrations, W. H. Matthews, *Mazes and Labyrinths – Their History and Development,* New York (Dover) 1970, etc.

27. Eden Project welcomes online visitors at http://www.edenproject.com/
28. Alden Hatch, *Buckminster Fuller – At Home in the Universe,* New York (Crown) 1974, pp. 214-15.
29. R. Buckminster Fuller, "Ten Proposals for Improving the World," in his *Earth, Inc., op. cit.,* 173-180.
30. Jawaharlal Nehru, *The Discovery of India,* Robert I. Crane, Ed., New York (John Day/Anchor) 1946/1960, pp. 210-11.
31. Cf. the entire article by Ashis Nandy, "Science, Authoritarianism and Culture," *INTERculture* #112 (Vol. XXIV, No. 3), Summer 1991, pp. 9-32 (citation from p. 20).
32. Cf. Sachs' Introduction to Wolfgang Sachs, Ed., *The Development Dictionary – A Guide to Knowledge as Power,* London (Zed Books) 1992, pp. 1-5.
33. J. Nehru, *The Discovery of India, op.cit.,* p. 31.
34. Cf., e.g., C. G. Jung, *Mandala Symbolism,* R. F. C. Hull, trans., Princeton (Bollingen) 1959, 1972. Jung's position on the role in such diagrams of the 'collective unconscious' is well known: "In view of the fact that all the mandalas shown here were new and uninfluenced products, we are driven to the conclusion that there must be a transconscious disposition in every individual which is able to produce the same or very similar symbols at all times and in all places. Since this disposition is usually not a conscious possession of the individual, I have called it the *collective unconscious,* and, as the bases of its symbolical products, I postulate the existence of primordial images, the *archetypes.*" (p. 100)
35. Cf., e.g., R. Panikkar, *Myth, Faith & Hermeneutics,* New York (Paulist Press) 1979, Introduction, especially pp. 6-8, for his notion of symbol.
36. Gregory Bateson, *Mind & Nature – A Necessary Unity,* New York (Dutton) 1979, p. 17.
37. Cf. Nigel Pennick, *Sacred Geometry,* New York (Dover) 1981, as well as Nigel Pennick, *The Ancient Science of Geomancy – Man in Harmony with the Earth,* London (Thames & Hudson) 1979: "Geometry is fundamental; it is universal. The structure of things, both organic and inorganic, is based upon the universal rules of geometry. Through geometry, the living and non-living, the natural and the artificial, are linked with the various material, psychological and spiritual planes which constitute the basis of religious experience. ... The geometrical basis of the temple is deliberately designed as a microcosmic image of universal order, acting as a catalyst which enables the individual to harmonise with the cosmos. Hence the temple itself must be an accurate reflection of the cosmos, and be in harmony with it." (from pp.116-18) Cf. also John Michell, *The View Over Atlantis,* New York (Ballantine) 1969; inartfully titled, but offering the profound symbol of the Earth as

a living organism with its own neural circuitry, reminiscent of Teilhard de Chardin's "noosphere."

38. Cf. the classic works of G. Tucci, *The Theory and Practice of the Mandala*, A. H. Brodrick, trans., New York (Samuel Weiser) 1969, C. G. Jung, *Mandala Symbolism, op. cit.*, and Mircea Eliade, *Cosmos and History*, W. R. Trask, trans., Princeton, NJ (Bollingen) 1954 and New York (Harper & Row) 1959 (especially Ch. One); as well as more popular pictorial works like J. Cornell, *Mandala*, Wheaton, IL (Quest) 1994, and of course J. & M. Argüelles, *Mandala*, Boulder & London (Shambhala) 1972: "The universality of the mandala is in its one constant, *the principle of the center.* (p. 12) ... It is the gastepost between the macrocosm and the microcosm." (p. 15)
39. Cf. also the bridging work of Roberts Avens, *Imagination Is Reality*, Dallas (Spring) 1980 for what Avens calls "Western Nirvana" in Jung, Hillman, Barfield and Cassirer: resouling the world.
40. For another twentieth-century effort to see the mandala of creation mirrored in the evolution of human awareness, cf. Pierre Teilhard de Chardin, *The Phenomenon of Man*, New York (Harper & Row) 1965; from the Foreword, "Seeing": "I repeat that my only aim, and my only vantage-ground in these pages, is to try to see; that is to say, to try to develop a *homogenous* and *coherent* perspective of our general extended experience of man. A *whole* which unfolds. ... Like the meridians as they approach the poles, science, philosophy and religion are bound to converge as they draw nearer to the whole. I say 'converge' advisedly, but without merging, and without ceasing, to the very end, to assail the real from different angles and on different planes." (p. 36)
41. R. Buckminster Fuller with E. J. Applewhite, *Synergetics 2*, London/New York (Collier/Macmillan) 1979, ¶1052.54-57, pp. 418-19.
42. See Appendix B, "The Root DHṚ and Its Branches."
43. Cf. Gladys A. Reichard, *Navajo Religion – A Study of Symbolism*, Princeton (Bollingen) 1950.
44. Cf., e.g., R. Buckminster Fuller, *Synergetics, op. cit.* ¶905.02, p. 487.
45. Cf. T. Roszak, M. E. Gomes, A. D. Kanner et al., *Ecopsychology*, San Francisco (Sierra Club) 1995.
46. Cf. Wesselow, "Thinking Hands," in *The Way of the Maker – Eric Wesselow's Life Through Art*, S. Eastham, Ed., New York/Oxford (University Press of America) 2001, pp. 9-12.
47. Snelson himself is famously bitter about Bucky's high-handed appropriation of what he terms his "floating compression" structures (Bucky called them "tensegrities"), the best known example of which is probably his great mast at the Smithsonian sculpture garden in Washington, D.C. He tells his own story online at: http://www.grunch.net/snelson//rmoto.html. Our experience of Bucky's generosity with his great-circle models, which we renamed WHOLES, is pretty much the opposite of Snelson's in every respect. See Appendix A, "Unfolding WHOLES."
48. E. Wesselow, *The Way of the Maker, op. cit.*, p. 43.
49. ibid., p. 29.
50. Cf. Amy Edmonson, *A Fuller Explanation – The Synergetic Geometry of R. Buckminster Fuller, op. cit.*, pp. 106-9.
51. Cf. R. Buckminster Fuller, *Synergetics, op. cit.*, ¶537.10-13, p. 297.
52. Cf. R. Buckminster Fuller, *Synergetics 2*, e.g., §1041.10, p. 406, "Seven Axes of Truncated Tetrahedron"; ¶986.303-4, p. 250; etc.
53. R. B. Fuller, *Synergetics, op. cit.*, ¶1022.10, "Minimum Sphere," pp. 654-5.
54. G. P. Flanagan, *Beyond Pyramid Power*, Marina del Rey, CA (DeVorss) 1975, pp. 16-20.
55. Cf. Peter Tompkins, *Mysteries of the Mexican Pyramids*, New York (Harper & Row) 1975, p. 281. Cf. also, e.g., Hugh Harleston, Jr., *Did Teotihuacan's designers, or their predecessors, have a knowledge of spherical trigonometry? : a research summary*, Mexico DF (UAC-KAN Research

Group) 1981, or his other works along these lines.

56. Cf., e.g., Ernest G. McLain, *The Myth of Invariance,* Boulder & London (Shambhala) 1978, p. 5: "The logic of India is profoundly geometric. Its mandalas and yantras present the observer with static forms which could only be achieved by dynamic processes. Our problem here is to learn to see those forms as Socrates yearned to see his own ideal forms, 'in motion' "; or E. A. S. Butterworth, *The Tree at the Navel of the Earth,* Berlin (Walter De Gruyter & Co.) 1970; or René Guénon, *The Symbolism of the Cross,* Paris (Gallimard) 1945 and London (Luzac & Co.) 1951, a gnostic vision of the cross which culminates in a ninefold spherical vortex; or the beautiful illustrations of Jill Purce, *The Mystic Spiral – Journey of the Soul,* New York (Avon) 1974.
57. For just one striking parallel, cf. C. G. Jung and R. Wilhelm, *The Secret of the Golden Flower,* New York (Harcourt, Brace & World) 1962, especially Jung's commentary, pp. 81-136: "Things [mandalas] reaching so far back in human history naturally touch upon the deepest layers of the unconscious and affect the latter where conscious speech shows itself to be quite impotent. Such things cannot be thought up but must grow again from the forgotten depths, if they are to express the deepest insights of consciousness and the loftiest intuitions of the spirit. Coming from these depths they blend together the uniqueness of present-day consciousness with the age-old past of life." For what may be the root intuition, cf. R. Panikkar, "Skambha – The Cosmic Pillar," in his *The Vedic Experience – An Anthology of the Vedas for Modern Man and Contemporary Celebration,* Berkeley/Los Angeles (University of California Press) 1977, pp. 61-67: "The intuition regarding this Cosmic Pillar or Support does not consist in seeing it, but in discovering the vestiges of its feet when they have disappeared in order to jump into the real; it is like seeing the vibrations of the springboard a moment after the dive. To know *skambha* is to know the Lord of Creatures without his creatures and without his Lordship." (p. 62). Cf. also Giorgio de Santillana and Hertha Von Dechend, *Hamlet's Mill,* Boston (Godine) 1977. Hamlet's Mill, or Amlodhi's Cairn, turns out after extensive investigations to be the Sampo of the *Kalevala,* i.e., a Hyperborean variant of *skambha,* the great central axis of the universe.
58. G. Tucci, *Theory and Practice of the Mandala, op. cit.,* Plate III.
59. For such 'octa voids' in Buddhism, cf. A. F. Price & Wong Muo-Lam, *The Diamond Sutra and the Sutra of Hui Neng,* Boulder (Shambhala) 1969.
60. M. Singh, *Himalayan Art,* New York (Macmillan) 1965.
61. "Last fall the United Nations Human Settlements Program published a historic report, "The Challenge of Slums," warning that slums across the world were growing in their own hothouse, viral fashion. One billion people, mainly uprooted rural migrants, are currently warehoused in shantytowns and squatters' camps, and the number will double in the next generation. The authors of the report broke with traditional UN circumspection to squarely blame the International Monetary Fund (IMF) and its neocolonial 'conditionalities' for spawning slums by decimating public sector spending and local manufacturing throughout the developing world. During the debt crisis of the 1980s, the IMF, backed by the Reagan and Bush administrations, forced most of the third world to downsize public employment, devalue currencies and open their domestic markets to imports. The results everywhere were an explosion of urban poverty and sharp fall-offs in public services," Mike Davis, "A Plague of Slums," *AlterNet,* 3 February 2004: http://www.alternet.org/story.html?StoryID=17735.
62. Cf. Jacques Ellul, *The Technological Bluff,* G. W. Bromiley, Trans. (Grand Rapids: Eerdmans, 1990), e.g., p.145: "The intellectual and cultural tragedy of the modern world is that we are in a technical milieu that does not allow reflection. We cannot look at the past and consider it. We cannot fix on an object and reflect on it. The technical object encompasses us even though we know nothing about it ..." Cf. also Scott Eastham, *Biotech Time-Bomb,*

Auckland (R S V P) 2003, especially "The Automation Idea," pp. 55-59.

63. Cf. Scott Eastham, "Via Media – The 'Future' of Media Studies," *New Zealand Journal of Media Studies,* Vol. 2, No. 2, December 1995, pp. 37-44.
64. Cf. Scott Eastham, "In Pound's China – The Stone Books Speak," in *Paideuma,* Vol. 31, Nos. 1 & 2, Spring & Fall, 2003, §III.1 "The Mythic Horizon."
65. Cf. Jacques Ellul, *Propaganda – The Formation of Men's Attitudes,* New York (Knopf) 1965.
66. Cf. B. R. Barber, "Jihad Vs. McWorld," in *The Atlantic Monthly,* March 1992, pp. 53-63.
67. For a picaresque but thought-provoking survey of these paradoxical effects of the new communications technologies, cf. Alexander Stille, *The Future of the Past – The Loss of Knowledge in the Age of Information,* London (Picador) 2003.
68. For these diverging intellectual vectors of the late twentieth century, compare Jürgen Habermas, *Knowledge & Human Interests,* J. J. Shapiro, trans., Boston (Beacon Press) 1971, and Hans-Georg Gadamer, *Truth & Method,* G. Barden & J. Cumming, Ed., New York (Crossroad) 1982.
69. John Ruskin, *Collected Works,* Vol. VI, p. 346, "Influence of Imagination on Architecture."
70. For this dynamic of *mythos* and *logos*, cf. R. Panikkar, *Myth, Faith & Hermeneutics, op. cit.*
71. Edward Goldsmith, *The Way – An Ecological World-View,* Foxhole, UK (Themis Books) 1996.
72. The same meaning turns up in the Old High German *twellan* and the Old Norse *dvelja,* to retard, delay; and is heightened in the Middle Dutch *dwellen,* to stun, make giddy, perplex.
73. For direct contrast, cf. Sidney Pobihushchy & David Bedford, "Towards a People's Economy: The Co-op Atlantic Experience," *INTERculture* #120, Vol. XXVI, No. 3, Summer 1993, pp. 2-42, focused not on the migrations of 'horse-people' but on workable local dwelling strategies.
74. *Encyclopædia Brittanica,* 1958 Edition, Vol. 11, p. 754c.
75. Cf. Lewis Mumford, *The Myth of the Machine,* Vol. 1, *Technics and Human Development, op. cit.,* p. 172.
76. Cf., e.g., R. Buckminster Fuller, "Twilight of the World's Power Structures," in *Critical Path, op. cit.,* pp xvii-xxxviii.
77. Cf. E. F. Schumacher, *Small is Beautiful – Economics as if People Mattered*, New York (Harper & Row) 1973, Part I, 4, "Buddhist Economics," pp. 53-62. For a simple but powerful recent condensation of Buddhist economics, cf. James W. Heisig, *Dialogues At One Inch Above the Ground – Reclamations of Belief in an Interreligious Age,* New York (Herder/Crossroad) 2003, "Sufficiency and Satisfaction: Recovering an Ancient Symbolon," pp. 5-29: "Some years ago I argued for the introduction of a 'principle of sufficiency' into economic theory and consumer ethics. My guiding image was a cryptic saying carved into a *tsukubai,* or stone water basin, at Ryoan-ji in Kyoto ... *All I know is how much is enough.*" (p. 7, emphasis mine)
78. Excerpted from R. B. Fuller, *Critical Path, op. cit.,* pp. 64-6.
79. Cf. R. Buckminster Fuller, *Grunch of Giants*, New York (St. Martin's) 1983.
80. Along these lines, it was at Norman Cousins' suggestion that Fuller produced the remarkably humble (for Bucky) poetic sequence "How Little I Know," in *It Came to Pass – Not to Stay,* London & New York (Collier/Macmillan) 1976, pp.1-56.
81. Cf. R. Panikkar, *Myth, Faith & Hermeneutics, op.cit.*, Part I, 'Myth.'
82. This is the title thematic of R. Panikkar, *The Cosmotheandric Experience – Emerging Religious Consciousness*, S. Eastham, Ed., New York (Orbis) 1993.
83. A characterization I owe to Ntsuk, 'Otter,' my 'sauvage' Montagnais colleague from the Intercultural Institute of Montréal in Québec.
84. Alan Ereira himself has written what must be the most poignant critique, "Back to the Heart of Lightness," in *The Ecologist*, Vol. 31, No. 6, July/Aug. 2001, pp. 34-38. Revisiting

the Kogi a decade after his remarkable 1990 BBC documentary – *From the Heart of the World: The Elder Brother's Warning* – he says: "That has been one consequence of the film. The Kogi have become a marketable commodity. ... Whether what is being sold is Kogi telepathy, Kogi coffee, or 'Kogi experience' trips into the jungle, they feel it all amounts to a theft of their public identity for profit, and one which would not have happened if the film had not been made." (p. 38) For the intriguing ninefold structure of the maternal matrix of the Kogi universe alluded to above, cf. Geraldo Reichel-Dolmatoff, "Funerary Customs and Religious Symbolism among the Kogi," in *Native South Americans: Ethnology of the Least Known Continent*, P. J. Lyon, Ed., Boston (Little Brown) 1974, and "Training for the Priesthood Among the Kogi of Columbia," in *Enculturation in Latin America: An Anthology*, J. Wilbert, Ed., UCLA Latin American Studies, Vol. 37, Los Angeles (UCLA Latin American Center Publications) 1976.

85. William McDonough, *Earth Island Journal*, Spring 1996, pp. 31-2.
86. R. Buckminster Fuller, *Critical Path, op. cit.*, pp. 123-160.
87. R. Buckminster Fuller, *Intuition, op. cit.*, pp. 89-97.
88. R. Buckminster Fuller, *Critical Path, op. cit.*, pp. 141-7
89. Cf. Scott Eastham, *EyeOpeners*, Wellington (Horizon Press) 1999, p. 181.
90. R. W. Emerson, "Brahma," *The Works of Ralph Waldo Emerson*, Vol. 4, *op.cit.*, p. 253.
91. R. Panikkar, *The Vedic Experience, op. cit.*, p. 58.
92. Cf. the works of Thomas Berry, e.g., *The Dream of the Earth*, San Francisco (Sierra Club Nature and Natural Philosophy Library) 1988; *The Great Work*, New York (Bell Tower) 1999, etc.
93. Cf. Buckminster Fuller & Anwar Dil, *Humans in Universe, op. cit.*
94. R. W. Emerson, *The Works of Ralph Waldo Emerson, op. cit.*, Vol. 4, pp. 1-55.
95. Cf. George Steiner, *Real Presences*, London (Faber) 1989.
96. R. W. Emerson, *The Works of Ralph Waldo Emerson*, Vol. 1, *op. cit.*, pp. 67-8.
97. R. W. Emerson, *The Works of Ralph Waldo Emerson*, Vol. 1, *op. cit.*, pp. 92-3.
98. See "Margaret Fuller's Prophecy," in R. B. Fuller, *Ideas and Integrities, op. cit.*, pp. 67-71.
99. Hugh Kenner, "Bubbles and Destiny" *Bucky, A Guided Tour of Buckminster Fuller, op. cit.*
100. Cf. C. G. Jung and Wolfgang Pauli, *The Interpretation of Nature and the Psyche*, Princeton (Bollingen/Pantheon) 1952, both for Pauli's contribution, "The Influence of Archetypal Ideas on the Scientific Theories of Kepler," and for Jung's famous essay, "Synchronicity, An Acausal Connecting Principle," from which: "Synchronicity is no more baffling or mysterious than the discontinuities of physics. It is only the ingrained belief in the sovereign power of causality that creates intellectual difficulties and makes it appear unthinkable that causeless events exist or could ever occur. But if they do, then we must regard them as *creative acts*, as the continuous creation of a pattern that exists from all eternity, repeats itself sporadically, and is not derivable from any known antecedents." (pp. 141-2)
101. Arthur Koestler, *Janus – A Summing Up, op. cit.*, pp. 269-70.
102. Cf. Robert D. Richardson, Jr., *Emerson – The Mind on Fire*, Berkeley/Los Angeles (University of California Press) 1995, for a resumé of when exactly Emerson was reading what.
103. Cf. Stanley Cavell, "Thinking of Emerson," in his *The Senses of Walden*, Chicago & London (University of Chicago Press) 1972, pp. 121-138; as well as his later *Conditions Handsome and Unhandsome – The Constitution of Emersonian Perfectionism* (Carus Lectures 1988), Chicago & London (Chicago University Press) 1990.
104. Cf. R. Panikkar, "The Supreme Experience: The Ways of West and East," in his *Myth, Faith & Hermeneutics, op. cit.*, Part 3.b, "Four Archetypes of the Ultimate," pp. 311-15.
105. Cf. Rudolph Otto, *The Idea of the Holy*, J. W. Harvey, trans., London (Oxford) 1923, 1969.
106. Again, cf. Panikkar's "Supreme Experience," in *Myth, Faith & Hermeneutics, op. cit., loc. cit.*

107. ibid., p. 313.
108. R. Buckminster Fuller, *Intuition, op. cit.*, p. 219 *et seq.*, "Two Versions of the Lord's Prayer."
109. Cf. Christopher Isherwood, *Ramakrishna and His Disciples,* London (Methuen & Co.) 1965.
110. Cf. R. Panikkar, "Human Rights, A Western Concept?," in *INTERculture* #82, Intercultural Institute of Montréal, Vol. XVII, Nos. 1-2, January-March 1984, pp. 28-47.
111. Cf. the Latin *tenere,* French *tenir* and English *tenet*. See also Appendix B: "The Root DHṚ and Its Branches."
112. R. Panikkar, "Human Rights, A Western Concept," *INTERculture* #82, *op.cit., loc. cit.*, p. 39.
113. Cf. Frederick J. Streng, *Emptiness – A Study in Religious Meaning,* Nashville/New York (Abingdon) 1967.
114. Cf. S. T. Stcherbatsky, *The Central Concept of Buddhism and the Meaning of the Word* 'Dharma,' Calcutta, 1961, as well as his better known *Buddhist Logic,* New York (Dover) 1964.
115. R. B. Fuller, *Synergetics, op. cit.* ¶700.04, p. 372.
116. Cf. Peter Brook's powerful, though rather secular, filmed version of the *Mahabharata* (1989) for a dramatization of this fateful scene where Natchiketas encounters Dharma 'in person'.
117. Back cover endorsement for Scott Eastham, *Paradise and Ezra Pound – The Poet as Shaman,* New York & London (University Press of America: paperback edition) 1983.

Appendix A: Unfolding Wholes

1. Fuller, *Synergetics, op. cit.*, ¶101.01 et seq. References unless otherwise noted are to the numbered paragraphs of *Synergetics* and *Synergetics* 2, R. B. Fuller & E. J. Applewhite, New York (Macmillan) 1975, 1979. "*Synergetics* is a book about models: humanly conceptual models; lucidly conceptual models; primitively simple models; rationally intertransforming models; and the primitively simple numbers uniquely and holistically identifying those models and their intertransformative, generalized and special case, number value accountings." (¶900.21)
2. Even thoughts disclose their own (tetrahedral) 'conceptual geometry,' as Fuller deftly explicates in the important essay "Omnidirectional Halo," in R. Buckminster Fuller, *No More Secondhand God,* New York, *op. cit.*, pp. 115-145: "Intellect may be 'creating,' finitely extending and re-fining universe as it asks each next good question." (p. 145)
3. Cf., e.g., R. Buckminster Fuller, *Utopia* or *Oblivion, op. cit.*, p. 95.
4. See Mary Laycock, *Bucky for Beginners,* Hayward, CA (Activity Resources Co.) 1984, for ways to make many of the basic synergetics models using handy household materials (cut-out paper forms, toothpicks, straws, tape, glue, pins, etc.), fun projects for youthful explorers with clear, easy-to-follow instructions.
5. E. J. Applewhite, *Cosmic Fishing, op. cit.*, p. 10. Cf. also R. Buckminster Fuller, *Synergetics, op. cit.*, ¶982.50, "Initial Four-Dimensional Modelability."
6. Amy Edmonson, *A Fuller Explanation, op. cit.*, p. 219.
7. ibid., pp. 220-221
8. Cf. Edwin Abbott Abbott, Flatland – *A Romance of Many Dimensions,* San Francisco (Arion) 1981, the late Victorian social satire on how a three-dimensional world would appear to two-dimensional beings; by extrapolation a critique of blockheadedness.
9. Cf. also R. Buckminster Fuller, *Synergetics, op. cit.*, §840.00, "Foldability of Four Great Circles of Vector Equilibrium."
10. E. J. Applewhite, *Cosmic Fishing, op. cit.*, p. 143.
11. Cf. P. D. Ouspensky, *A New Model of the Universe,* London (Arkana) 1984.
12. R. Marx, *Dymaxion World of R. Buckminster Fuller, op. cit.*, p. 42.
13. Hugh Kenner, *Geodesic Math, And How to Use It,* Berkeley/Los Angeles (University of

California Press) 1976, recently reprinted by popular demand.

14. A paper on the carbon-60 transistor projects, led by P. L. McEuen and A. P. Alivisatos at Lawrence Berkeley Lab, and H. Park of Harvard, appeared in *Nature*, Vol. 407, No. 6800, 7 September 2000, pp. 57-60.
15. There's a story behind this figure, and further exploration for the interested reader. When we tried to construct the rhombic triacontahedron from the specifications originally published in *Synergetics*, it just wouldn't work. Every which way we tried it, it ripped and tore ... which is fairly exasperating after you've cut, scored and folded up thirty little rhombic bow-ties. We contacted Fuller, who took the criticism in his stride and included a revised set of specifications (reproduced here) in *Synergetics 2*. Now this one does work, it can be assembled, but it seems to us that one angle is slightly 'off' and the result is a figure which very nearly splits open at the unpinned (central pentagonal and hexagonal) vertices. This may in fact be what Fuller was after in what he called his "Demass model," a figure which almost 'bursts' yet still holds together. Fuller passed away before we had a chance to show him a re-constructed figure with that one angle rectified (fairly simple to do). Our agreement with him in 1981 stipulated that we could reproduce drawings from *Synergetics* "as long as the data in them is in no way altered." In producing this Appendix, we must stick by our agreement, but might encourage the reader to experiment a little with the specifications for this figure. As Fuller used to say, in the faintly amused, slightly rueful tone of someone who knows all about trial-and-error: "You can never learn less."
16. Cf. R. B. Fuller, *Synergetics, op. cit.*, ¶1007.27.
17. Cf. Anthony Pugh, *An Introduction to Tensegrity*, Berkeley/Los Angeles (University of California Press) 1976, and Hugh Kenner, *Geodesic Math, And How to Use It, op.cit.*

Index